Understanding
The Dairy Cow

Second Edition

John Webster
MA, VetMB, PhD, MRCVS

Professor of Animal Husbandry
University of Bristol
School of Veterinary Science

OXFORD

BLACKWELL SCIENTIFIC PUBLICATIONS

LONDON EDINBURGH BOSTON

MELBOURNE PARIS BERLIN VIENNA

Copyright © John Webster 1987, 1993

Blackwell Scientific Publications
Editorial Offices:
Osney Mead, Oxford OX2 0EL
25 John Street, London WC1N 2BL
23 Ainslie Place, Edinburgh EH3 6AJ
238 Main Street, Cambridge,
    Massachusetts 02142, USA
54 University Street, Carlton
    Victoria 3053, Australia

Other Editorial Offices:
Librairie Arnette SA
2, rue Casimir-Delavigne
76006 Paris
France

Blackwell Wissenschafts-Verlag GmbH
Meinekestrasse 4
D-1000 Berlin 15
Germany

Blackwell MZV
Feldgasse 13
A-1238 Wien
Austria

First published 1987
Second edition 1993

Set by DP Photosetting, Aylesbury,
Bucks
Printed and bound in Great Britain by
Hartnolls Ltd, Bodmin, Cornwall

DISTRIBUTORS

Marston Book Services Ltd
PO Box 87
Oxford OX2 0DT
(*Orders:* Tel: 0865 791155
    Fax: 0865 791927
    Telex: 837515)

USA
    Blackwell Scientific Publications, Inc.
    238 Main Street
    Cambridge, MA 02142
    (*Orders:* Tel: 800 759-6102
        617 876-7000)

Canada
    Oxford University Press
    70 Wynford Drive
    Don Mills
    Ontario M3C 1J9
    (*Orders:* Tel: 416 441-2941)

Australia
    Blackwell Scientific Publications Pty Ltd
    54 University Street
    Carlton, Victoria 3053
    (*Orders:* Tel: 03 347-5552)

British Library
Cataloguing in Publication Data
A Catalogue record for this book is
available from the British Library

ISBN 0-632-03438-6

Library of Congress
Cataloging in Publication Data
Webster, John, 1938
    Understanding the dairy cow/
    John Webster. — 2nd ed.
        p.   cm.
    Includes bibliographical references
and index.
    ISBN 0-632-03438-6
    1. Dairy cattle.   2. Cows.   I. Title.
    SF208.W4   1993   93-9146
    636.2'142—dc20   CIP

To the memory of Fred Robinson, herdsman, of Shutlanger, who understood dairy cows, and to Teddy Cook, veterinary surgeon, who introduced me to him and to so much else.

Moreover:
'The cow, crunching with depressed head, surpasses any statue .....
they are so placid and self-contained, I stand and look at them long
and long.'

Walt Whitman
'Song of Myself'

Contents

Acknowledgements

I thank the following for permission to reproduce illustrations: Geoffrey Pearson, Fig. 2.3; The Scottish Farm Buildings Investigation Unit, Fig. 8.1; Jim Pinsent, Fig. 8.7; The Institute for Food Research, Reading, Figs. 8.1 and 8.2; *The Veterinary Record*, Fig. 8.6.

I am extremely grateful to Geoff Davies, Susan Long, Norman Todd and Christopher Wathes for their criticism and advice on various chapters. Above all, my thanks are due to Pauline Webber who typed the manuscript and to Nick Jeanes who transformed my amateur sketches into such excellent, professional illustrations.

Preface to the Second Edition

Once upon a time the objectives of dairy farming were simple: to produce wholesome milk in large quantities and as cheaply as possible. Recently, however, and especially since this book was first published in 1987, dairy farmers and their advisers have been subjected to a barrage of new and conflicting incentives and constraints. On the one hand quotas, pollution controls and food scares restrict the amount of milk they can sell off the farm. On the other, drugs such as BST (bovine somatotropin) tempt them to increase output from the individual cow. To accommodate these constant and conflicting pressures, dairy farming needs to be flexible. This book is therefore not an instruction manual on dairy cow husbandry but an examination of the fundamental principles of physiology, health and behaviour that determine both efficient production of the dairy cow and proper attention to her welfare in any circumstances intensive or extensive, high-tech or 'organic'.

As before, the second edition is divided into four main sections: how the cow works, feeding, housing, health and management, and breeding and fertility. The introduction develops the theme of the role of animals (especially the dairy cow) in sustainable agricultural systems. The section on feeding contains new material on buffer feeding, whole-crop cereals and feeding by-products as 'straights'. It also contains a complete explanation of the new 'Metabolizable Protein' system adopted as the UK standard in 1992. New problems associated with BST and BSE (bovine spongiform encephalopathy or 'mad cow disease') are explained. A new chapter on breeding critically examines the practice, principles and limits of genetic improvement by multiple ovulation, embryo transfer and cloning. The first edition closed with a Coda that speculated on the future for the dairy cow. The new Coda reviews these prophecies to date then looks forward to a spectrum of future options for dairy cow husbandry ranging from the use of robotics to allow cows almost complete control of their own environment to the use of intermediate technology to roll back the major constraints on dairy production in the tropics.

Part I

How the Cow Works

1 Introduction – The Dairy Cow of Today

Understanding the dairy cow is a matter of heart and mind. It is essential to examine her scientifically as a complex and elegant machine for the production of milk, the nearest thing in nature to a complete food. It is equally essential to recognise her as a sentient (and highly engaging) creature with rights to a reasonable standard of living and a gentle death. In both senses of the word this understanding is not static. The more we study the workings of the dairy cow, the more efficiently we can exploit her capacity to produce milk and, eventually, meat, not only from grass and cereals but from an enormously wide range of plant products and by-products which we cannot, or choose not to, eat ourselves. The more we study her health, behaviour and environmental requirements the better we can ensure her welfare.

Consider the two pictures in Fig. 1.1. Both are of dairy cows. However, to borrow an aphorism from Professor Colin Spedding, both are, in fact, only pictures of the right side of the outside of dairy cows, i.e. they conceal much more than they reveal. The Dairy Shorthorn and Friesian cows respectively typify the size and shape deemed ideal for dairy production eighty years ago and now. Relative to the Shorthorn, the Friesian is heavier, rather more wedge-shaped, and has a larger udder. The larger udder produces more milk per cow and the larger, wedge-shaped animal eats more. Animal breeders achieve effects like these over a remarkably short period of time simply by selection for the desired trait (production of milk solids per animal), without needing to ask any questions as to what has taken place inside the cow to bring this about. However, when one starts to pose more subtle questions such as 'Was the increase in milk production accompanied by an increase in food conversion efficiency or an increased predisposition to production disease?', or 'Would a similar increase have been achieved with other foods or in a different environment?', then it becomes increasingly unsatisfactory to think only in terms of inputs and outputs entering and leaving what in

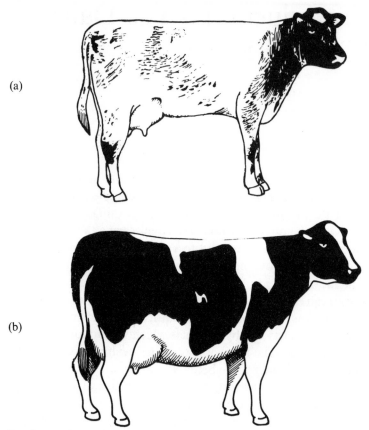

(a)

(b)

Fig. 1.1 (a) Young Dairy Shorthorn and (b) Friesian cows, animals considered to be of ideal dairy type in 1935 and 1985.

scientific jargon usually tends to get called a black box – or in this case a black and white box.

Most of the many books and articles on the dairy cow deal with the management of land, purchased food and the animals themselves to produce wholesome milk from healthy animals at least cost. Modern, concise books such as *Milk Production, Science and Practice* by Leaver (1983) or *Modern Milk Production* by Castle and Watkins (1984) are very good on how to feed and how to breed and these may be constantly augmented and brought up to date by booklets from advisory services: in the UK, the Agricultural Development and Advisory Service and the Milk Marketing Board. Practical books and articles like these tell the farmer or student what to do; how to manipulate the inputs to his black and white box in order to achieve outputs consistent with sound economics.

This is not one of those books, nor is it intended to be in competition with them. What I have attempted is a book for those such as students, stockmen, farmers and food compounders who see or read of a particular practical aspect of dairy farming and think 'But why?' or 'What if ...?'. These questions are perhaps provoked most often by brief reports in the farming press on allegedly new or improved methods. Articles like 'Flat-rate feeding brings dividends' or 'It's fodder-beet for me' are usually supported by figures to show that these practices have undoubtedly worked well on a particular farm. What the farmer reading the article needs to know is why did flat-rate feeding work better for Smith than for Jones (or better in 1992 than in 1993) and what would happen if he adopted the practice on his own farm.

If he really wants the answer to that he has no option but to take a deep breath and plunge into the principles of digestive physiology and nutrition of the ruminant. This poses a new set of problems. Most textbooks that deal with the workings of the cow, i.e. the physiology of digestion, reproduction, lactation, etc., are not written for farmers but for students of veterinary or animal science and usually by those (such as myself) who have spent much or all of their working lives in academic research. This introduces an element of complexity which may be necessary for the student working towards an Honours degree or actively engaged in research, but which tends to tell the farmer or student (or the animal scientist working in another speciality) rather more than he actually wants to know – in this case the answer to the question, 'Why does flat-rate feeding work in some cases and not in others?'.

Textbooks of animal science also tend to be filled with information that is vital to the fundamental or applied scientist but which may provoke from the farmer or non-specialist student the question, 'So what?'. It is of interest to the microbiologist, for example, to identify the hundreds of different species of microorganism to be found within the rumen, and fundamental information of this sort has been applied, for example, to the development of food additives. It is of interest to the reproductive physiologist to investigate in ever greater depth the cascade of hormones that regulate sexual function, and such research has led to major innovations in the practical control of reproduction. However, knowing the names and specific chemistry of, for example, all the individual microorganisms in the rumen or all the specific hormones involved in reproduction is of little practical use to the dairy farmer unless there is something he can do to manipulate them to advantage for himself and his animals.

This book is written for farmers, stockmen, students, advisors and others directly concerned with the business of dairy production and so concentrates on those aspects of function in health and disease that are (a) of economic or welfare importance and (b) capable of manipulation on the farm.

Part I, 'How the cow works', deals with the physiology of digestion, metabolism, reproduction and lactation, responses to the environment and behaviour. Although physiologists might find this section rather simplistic it is undoubtedly more complex than most textbooks on husbandry. However, it assumes no more than a good general knowledge of biology and chemistry and is, I hope, self-explanatory thereafter. Recommendations for further reading are given at the end of the book.

Parts II-IV deal respectively with feeding, housing and health, and breeding and fertility. In each case, husbandry practices are evaluated critically in the context of the physiology and behaviour of the cow. This is not an instruction manual which says 'Do this' and 'Do that'. It assumes that the reader has a reasonable practical knowledge of dairy cow management and so addresses the question 'When I do this, or if I do this, how will the cow respond?'.

Part V looks to the future and the role of the dairy cow in the 21st century as she and farming systems adapt to demands of market forces and society. Since Exodus these have usually tended to involve wide fluctuations between the years of poor and plenty, the years of the lean and the fat kine, when the farmer is alternately the jolly, red-faced uncle or the grasping, avaricious bespoiler of the environment, and milk is either the best food for building bonny children or a major cause of heart failure.

One can, with certainty, say little more of politics, fashion and market forces than that they will continue to fluctuate. However, the yield of nutrients from the land and the conversion of nutrients by cattle into milk and meat are governed by fundamental, logical and essentially invariable laws of biology. Some of these laws are complex and we do not yet fully comprehend any of them, but the more that we who are interested in dairy cows and dairy farming can understand the better we shall be able to preserve and develop our craft to meet both the needs of the passing moment and our long-term responsibilities as custodians of the land and the livestock.

Milk as a food for man

The quality of any food depends on its composition, described in terms of the nutrients it contains relative to the nutrient requirements of the animal that consumes that food. All animals, including man, require nutrients principally as a source of energy do to the work of the body.

The energy value of food, which used to be measured in calories but is now conventionally measured in joules (1 calorie = 4.2 joules), corresponds to the amount of heat liberated when that food is combusted completely in an instrument called a bomb calorimeter. One kilogram of wheat, for example, would generate about 16.6 Megajoules (MJ) of heat on combustion. Animals extract what energy they can from food by the processes of digestion, metabolise it and produce heat in the process. After energy, the principal nutrients required, in decreasing amounts, are amino acids from proteins, minerals and vitamins.

The daily requirements of an adult man and a 4-year-old growing child for the major and some of the minor nutrients are listed in Table 1.1. Let us suppose, for the sake of extreme simplicity, that this man and boy subsisted on a diet of breakfast cereal and 500 ml milk per day, which is equivalent to about 60 g of milk solids. Breakfast cereal contains about 1360 kJ and 13 g protein per 100 g (it says so on the packet). Combining this information with that in Table 1.1, we can calculate how much breakfast cereal needs to be eaten by man and boy (Table 1.2).

This very simple calculation and the more detailed information in Table 1.1 emphasise some very important points which tend to be overlooked in many popular articles on human nutrition and dietetics.

On this ration, the man and boy require, respectively, 860 g and 410 g of dry food per day. Requirements for the second most important nutrient, protein, are 87 g/day and 56 g/day, i.e. only 10% and 14% of the diet respectively. Absolute daily requirements for minerals and vitamins are much smaller still. In other words, an adult man requires about 90% of his food as a source of energy, the growing child about 85%. The carbohydrate (lactose) and butterfat in milk are there to provide energy and only to provide energy but this, quantitatively, is overwhelmingly the most important nutrient.

A balanced diet is one in which the nutrients match the specific requirements of the consumer. Man cannot live on bread alone (not quite) because it lacks some of the minor nutrients. However, as indicated above, milk and cereal can meet not only the energy but also

Table 1.1   Contribution of 0.5 l cow's milk per day to the nutrient requirements of a 4-year-old child and an adult man doing moderate work

	4-year-old child		Adult man	
	Daily requirement	% from 0.5 l milk	Daily requirement	%from 0.5 l milk
Energy (kilojoules)	6400 kJ	25	12600 kJ	13
Protein (grams)	56.0 g	30	87.0 g	20
Calcium	1.0 g	60	0.8 g	75
Iron	7.5 mg	2	12.0 mg	1
Vitamin A	3000 iu	30	5000 iu	15
Vitamin D	400 iu	2	–	–
Vitamin C	15.0 mg	70	20.0 mg	50
Vitamin B	0.6 mg	35	1.2 mg	20
nicotinic acid	6.0 mg	7	12.0 mg	3
riboflavin	0.9 mg	85	1.8 mg	45

Values taken from Kon and Cowie (1961)

Table 1.2   Meeting the energy and protein requirements of man and boy from a diet of milk and breakfast cereal

	Man	Boy
Daily requirement: energy (kJ)	12600	6400
protein (g)	87	56
Energy supplied by milk (60 g solids, kJ)	1600	1600
Energy required from cereal (kJ)	11000	4800
Cereal intake (g/d)	800	350
Protein yield from milk and cereal (g/d)	120	62
Protein balance (g/d)	+33	+6

the protein requirements even of growing children. When one comes to consider the minor nutrients the quality of milk as a food becomes even more impressive. Relative to its capacity to supply energy, milk is particularly rich in high quality protein, calcium, vitamin A and vitamin C. The only nutrients it lacks are iron, vitamin D and some of the B vitamins, e.g. nicotinic acid. In practice this only constitutes a problem for animals such as veal calves reared on milk alone. Man almost invariably gets an ample supply of iron and B vitamins from a wide range of natural foods. Vitamin D is less widely distributed in nature. Animal oils, classically fish liver oils, and egg yolks are rich in vitamin D. The vegetables that we commonly eat are, on the whole, relatively deficient in vitamin D but processed cereals are often 'fortified' with added vitamins (once again, it pays to read the packet).

As a food for man, therefore, cow's milk:

(1) is rich in almost all the nutrients essential for life.
(2) relative to energy, is particularly rich in essential nutrients such as high-quality protein, calcium and vitamin A so can be used to balance diets based on cereals or root crops which are deficient in these things.
(3) can be processed into a wide range of foods such as cheese, butter, yoghurt, etc., which not only add variety of taste but also vary the composition of the food so as to improve the overall balance of the daily diet according to what else is being eaten and how much.

In recent years, the image of milk as the nearest thing in nature to a perfect food has come under attack from two directions at once. One is the vegetarian/vegan case that man has either limited or no right to exploit animals for food; the other is that milk can be positively unhealthy. These two arguments are obviously different although frequently cobbled together by those wishing to make a point. I shall consider the vegetarian case in the next section (The biological efficiency of milk production) and elsewhere. The associations between the nutrients contained in wholesome milk and human health are complex. Very briefly:

Cow's milk and human milk both contain carbohydrate as the disaccharide, lactose (see Table 1.6 later in this chapter). Newborn infants possess lactase, the enzyme that breaks down lactose into glucose and galactose. However, in many parts of the world such as South East Asia, children lose the ability to manufacture lactase after weaning off their mothers and develop an intolerance to lactose in whole milk. However, they retain the ability to digest milk fat and protein in butter and cheese, and can also eat yoghurt and ghee since the lactose has been fermented to lactic acid. It is equally the case that certain individuals of all races develop allergies to proteins in cow's milk. Obviously the ideal food for a young baby is milk from its own mother both as a source of nutrients and a continuing source of antibodies to protect the epithelial surface of the gut. When a baby first experiences cow's milk or any other novel source of protein, its immune system must first identify it and then recognise it as harmless. Normally, immunological recognition of new food proteins proceeds to tolerance. Occasionally, perhaps when the new protein is presented to the gut at the same time as a harmful antigen (e.g. from a virus or bacteria), recognition may proceed to hypersensitivity and the development of food allergy. This is largely a matter of bad luck and cow's milk cannot be condemned as the guilty party. A child that acquires an allergy to cow's milk in early life may thrive on goat's milk. By the same logic, if it had been drinking goat's milk when its immune system was tricked into hypersensitivity, it would be likely to thrive on cow's milk.

Approximately 70% of the fatty acids in milk are fully saturated. The remaining 30% is mainly monosaturated oleic acid (18:1). The incidence of coronary heart disease has been linked to the intake of saturated fats (as distinct from overeating *per se*) by the UK Committee on Medical Aspects of Food Policy (1984). However, this conclusion which was based on a selected sample of statistics from only seven nations when 21 were available, has become increasingly untenable in the light of more recent evidence (reviewed by Blaxter and Webster, 1991). Moreover, oleic acid, the principal oil in the healthy Mediterranean diet *may* be positively beneficial though we cannot be sure. The only responsible way to summarise the huge volume of evidence on diet and coronary heart disease is to conclude that overeating is dangerous but that there is no longer justification for generating food scares based on the chemical composition of fat in milk, or indeed meat.

Biological efficiency of milk production

Cows harvest grass and a wide range of other plant materials to produce milk. The source of energy for all natural food production is, of course, the sun. The theoretical maximum efficiency of capture of solar energy by green plants is about 3%. Actual efficiencies of conversion of solar energy to food for animals and man are all below 1% (Table 1.3). The low efficiency of capture of solar energy is not too alarming as there is a lot of it about and it will be around for some time yet! Grass is the most efficient of the crops listed because it can be harvested more than once a year and the whole crop gets eaten. When man grows cereals or other crops such as potatoes for his own consumption he usually grows only one crop a year and inevitably a proportion of the crop is discarded.

Table 1.3 Efficiency of conversion of solar energy into chemical energy in food

	Yield	Efficiency of solar capture (%)
Grass	50 tonnes/ha	0.71
Wheat grain	7 tonnes/ha	0.34
Potatoes	35 tonnes/ha	0.50
Milk	25 litres/day	0.15
Pork	30 g 'meat'/day	0.04
Eggs	1 egg/day	0.06

Milk production is an efficient way of converting solar energy into food of animal origin because ruminants eat whole crops not directly available to man, we do not have to sacrifice them to obtain the milk and we can drink all the milk – there are no inedible parts like bones and offals.

Table 1.4 explores the efficiency of production of milk, meat and eggs relative to inputs of energy (and protein) as animal feed and energy as fossil fuel. Much of the information is derived from *Food, Energy and Society* by Pimentel and Pimentel (1979). In this table the costs relate not only to the requirements of the productive animals, e.g. the lactating dairy cow or laying hen, but also the support animals, e.g. maiden heifers, breeding sows and suckler cows. When all these factors are taken into account the intensive production of eggs and pork is quite similar to dairy production in the efficiency of conversion of animal feed into food for man partly because of the low proportion of food consumed by the 'support' animals. Beef production from suckler cows on extensive range is very inefficient because of the low yield relative to the costs of maintaining the breeding population.

At present all forms of animal production in developed countries consume more energy from fossil fuels than they generate in the form of food energy for man. In this sense all forms of animal production (and most other forms of agriculture) are ultimately unsustainable. Milk production as currently practised in the UK is particularly profligate with fossil fuels, consumed to produce machinery, refrigeration and, especially, fertilizers. I shall return to this dilemma in the concluding chapter. Extensive systems for the production of beef or

Table 1.4 Energy use for milk, meat and egg production from one breeding female per year

	Hen	Sow	Dairy cow	Range cow
Primary yield/year	300 eggs	1300 kg pork	6000 l milk	290 kg beef
Allocation of feed energy (kJ/MJ)				
to productive animals	790	830	720	580
to 'support' animals	210	170	280	420
Efficiency of yield				
kJ/MJ feed energy	140	182	170	37
g protein/MJ feed energy	2.31	1.09	1.70	0.16
kJ/MJ fuel energy	210	265	136	550
MJ/MJ food for man[1]	0.25	0.35	1.76	0.35
g protein/g protein food for man[1]	0.37	0.19	1.6	0.14

[1] Maximum yield of food energy and protein relative to intake of food which could have been fed directly to man

lamb consume relatively small amounts of fossil fuels but because productivity is so low they still only generate about half the energy as food that they consume as fossil fuel.

Perhaps the most ecologically attractive feature of the dairy cow is her ability to generate high quality food for man using resources which we cannot or choose not to eat ourselves. Modern rations for dairy cows may contain only 15% of their energy in the form of cereals or similar foods directly available to man. Pigs and poultry require at least 75–80% of their food in a form which we could eat ourselves. Table 1.4 shows that the dairy cow can generate a net gain of over 70% in terms of food for man whether expressed in terms of energy or protein.

There is no doubt that in the last 40 years the affluent have fed to animals a lot of food that could have been used to feed the poor. There are important moral issues here which extend beyond the scope of this book. The central, dominant economic issue has been that the poor in the Third World cannot afford to buy the cereals which we can produce in excess of our immediate needs. The most important biological fact which I wish to stress at this stage is that production of meat, milk and eggs may have been but does not have to be parasitic on the food needs of man. Dairying can be supported almost entirely from complementary food sources. Pigs and poultry eat food similar to ourselves and were most ecologically efficient when they scavenged the food we dropped or threw away. They can become equally efficient in the supermarket society when they scavenge the food that has passed its 'sell-by' date. The only food strategy for man that guarantees waste is vegetarianism.

Let us examine now in a little more detail the steps involved in converting the food eaten by animals (including man, or in this case woman) to milk, and thereby introduce a number of important terms which will recur in the sections on physiology, nutrition and practical feeding.

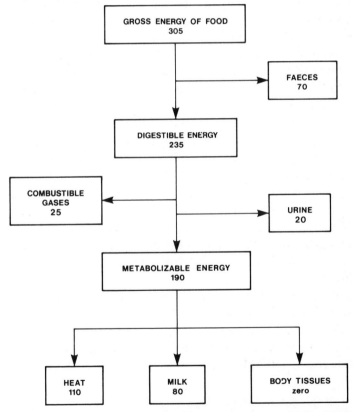

Fig. 1.2 Daily utilisation of food energy (MJ) by a 600 kg cow yielding 25 l/day of milk, eating a conventional diet of grass silage and dairy cake and neither gaining nor losing weight.

Figure 1.2 examines the fate of the energy contained in a typical meal of 8 kg cattle cake and 40 kg grass silage (24% dry matter) eaten by a typical 600 kg Friesian cow with a typical yield of 25 l/day.

(1) Total intake of *gross energy* is 305 MJ/day.
(2) *Faecal energy* losses account for 70 MJ/day. The amount that has apparently been digested and absorbed, *digestible energy* (strictly speaking, *apparent digestible energy*), is 235 MJ/day. The *apparent digestibility* of dietary energy is 235/305 = 0.77 (77%).
(3) Fermentation of carbohydrate foods in the anaerobic environment of the rumen generates considerable amounts of methane and other *combustible gases* which constitute a further loss of energy to the animal, in this case 25 MJ.
(4) Not all energy that is digested and absorbed by the animal can be combusted completely to carbon dioxide and water. Incompletely oxidised substances such as urea are excreted (mainly) via the urine and constitute a loss of *urine energy*, here 20 MJ.
(5) The energy remaining after deduction of losses in faeces, urine and as methane is called *metabolizable energy* (ME). Once again, this actually describes *apparently metabolizable energy* (see Chapter 5). ME may be considered in biological terms to be a measure of the capacity of a food or food mixture to provide fuel to do the work of the body. In more practical terms ME is now the standard method of feed description for cattle feeding in the United Kingdom and variants on the ME system of feeding are in use all over the world. The *metabolizability* of the gross energy in the diet (usually abbreviated as q) is 190/305 = 0.62.
(6) Metabolizable energy is either used for work and ultimately dissipated as heat, stored in the body tissues or secreted, in this case, as milk. The energy value of the milk is thus 80/25 = 3.2 MJ/ litre.
(7) This cow is neither gaining nor losing body tissue. ME is not being stored nor is the cow using her own body energy reserves to produce milk, 'milking off her back'. All ME not converted to milk appears therefore as heat. The *heat production* or *metabolic rate* of this cow is 110 MJ/day.

Of the 305 MJ of gross energy:

115 MJ of combustible energy has been lost in excreta (faeces, urine and methane)

110 MJ has been lost as heat (H)

80 MJ has been 'captured' as milk (RE_m)

Metabolizability

The ability of an animal to derive ME from the food it eats is determined by the chemical composition of the food and the digestive system of the animal that eats it. Man, for example, digests proteins, fats and sugars well, starches less well unless they have been cooked, and the fibrous material of plant cell walls (cellulose, hemicellulose and lignin) hardly at all. It would have been very difficult for a human to have been a vegetarian before learning to cook. The pig is rather more omnivorous than the human because it digests uncooked starches much better and, having a larger fermentation apparatus in the hind gut, can extract a little more nutrient from fibre. The ruminant is a true herbivore, having large fermentation vats for the digestion of plant fibre at each end of the gut. The effect of increasing the proportion of plant cell wall fibre on the metabolizability of food for man, pigs and cattle is illustrated in Fig. 1.3. Given a food with little or no plant fibre, ruminants are less efficient than pigs or man in obtaining ME from food, principally because of the energy losses in combustible gases.

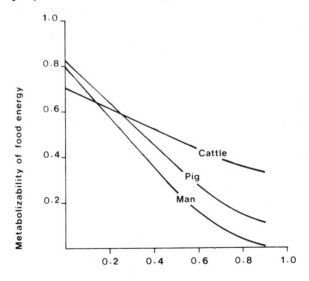

Fig. 1.3 Effect of increasing the proportion of plant cell wall fibre on the metabolizability of diets for cattle, pigs and man.

However, as the plant fibre content increases, so too does the relative efficiency of the ruminant.

This applies not only to fibre in grass but also that contained in by-products of food for man such as grain offals ('middlings', bran), soya bean meal, sugar beet pulp, maize gluten and crop residues such as straw. Table 1.5 describes the composition of a typical winter ration for a modern dairy cow. The ration contains 8 kg dry matter as silage, 2 kg as fodder beet, both grown on farm plus a further 8 kg of purchased dairy concentrate (7.3 kg dry matter) which provides 45% of total ME intake. However, this dairy cake, for sound economic reasons, contains relatively small quantities of 'straight' cereals like barley and wheat; most of the energy comes from by-products. The dairy cow does not therefore have to subsist on grass to avoid competing with man.

Utilisation of metabolizable energy

The efficiency with which a lactating mammal can convert ME and other absorbed nutrients into milk depends on its physiological

Table 1.5   Typical winter ration for a dairy cow yielding 25 litres of milk per day

	Fresh weight (kg)	Dry matter (kg)	Metabolizable energy (MJ)
Grass silage	32	8.0	80
Fodder beet	10	2.0	25
Dairy compound, total	8	7.3	95
'straight' cereals	2.4	2.2	20
by-products, cereals	3.2	2.9	38
oilseed meals	1.6	1.5	20
fish meal	0.5	0.4	7
minerals and vitamins	0.3	0.3	nil
Total	50	17.3	210
Contribution from 'straight' cereals (%)	6.4	13	14

capacity to synthesise milk and on the amount of ME and other nutrients required to sustain other essential functions. These are called *maintenance requirements*. An adult animal at rest and neither pregnant nor lactating is at maintenance when daily intake of nutrients exactly matches daily loss. In the case of energy metabolism, this occurs when ME intake equals heat production. Maintenance requirement for ME is primarily a function of body size (a cow requires more food energy than a chicken). Increments of ME in excess of maintenance requirement are used with an efficiency less than 1.0 to support milk synthesis. In the adult, non-pregnant, non-lactating animal, increments of ME in excess of maintenance requirement are deposited as fat again at an efficiency less than 1.0. In other words, the more an animal of any size eats, the more heat it produces (for details, see Chapter 2). The increase in heat production (H) per unit increase in ME intake ($kJ, H/MJ, ME$) is called the *heat increment of feeding*. In short, for any animal at rest, H is related to body weight ($f_1 W$) and food intake ($f_2 W$):

$$H = f_1 W + f_2 ME$$

The overall efficiency of utilisation of ME for milk production is determined therefore by the partition of ME between milk, body tissues and heat production.

Metabolic body size

It is obviously essential to account for the effects of differences in body weight (W) on H when comparing the efficiency of milk production, whether between species like cows and goats or between breeds of cattle like Friesians and Jerseys.

The relationship between ME requirement for maintenance (ME_m) and body weight is curvilinear, i.e. ME_m *expressed per kg body weight* is smaller in large animals than in small ones. In order to simplify this curve into a straight line, we convert both ME_m and H to $\log_{10}$ (Fig. 1.4). This gives a linear relationship with a slope of 0.75. Thus:

$$\log ME_m = 2.6 + 0.75 \log W$$

Reverting to real numbers:

$$ME_m = 400 \, W^{0.75}$$

Putting this into words, the ME requirement for maintenance of any

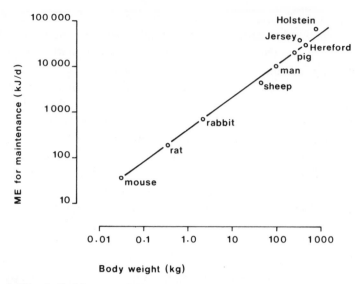

Fig. 1.4 Metabolisable energy (ME) requirements for maintenance in different species and different breeds of cattle plotted against body weight using a logarithmic scale for both axes.

mammal at rest is, as a first approximation, 400 kJ per kg body weight to the power 0.75. In biological terms this reflects the fact that ME_m and H are related more closely to body surface area than to body weight. In practical terms it means that we can not only account for effects of W or H when comparing animals of different size but can also calculate approximately how much food any mammal (or bird) needs for maintenance. The term $W^{0.75}$ is called *metabolic body size* (Kleiber, 1961); the exponent 0.75 is that which confers proportionality on measurements made of maintenance requirement in homeotherms (mammals and birds) differing in body weight.

Species and breed comparisons

Table 1.6 compares the yields and composition of milk from a variety of mammals. Milk from a Friesian cow is richer in protein and minerals than human breast milk, has a similar fat content but less lactose. The milk from a British Saanen goat is very similar to that of a Friesian cow in terms of the main chemical constituents (although the physical form differs – goat's milk appears more homogenous). Sows and bitches who give birth to large litters of relatively immature offspring secrete relatively rich milk. The most intriguing point made

Table 1.6 Yield and composition of milk by different mammals

	Friesian cow	Jersey cow	Saanen goat	Sow	Bitch	Woman
Body weight (kg)	600	400	65	200	26	60
Metabolic weight (kg$^{0.75}$)	121	89	22.9	53	11.5	21.5
Weight (no.) of offspring at birth(kg)	45(1)	30(1)	5(1)	16(12)	2.0(8)	3.5(1)
Peak milk yield (l/d)	31	21	5.0	7.5	1.3	1.0
Composition of milk (g/l)						
protein	33	37	35	60	83	12
fat	37	49	42	83	97	38
lactose	45	46	43	52	41	70
calcium	1.2	1.4	1.4	2.7	3.0	0.3
phosphorus	0.9	1.0	1.2	1.6	2.0	0.2
Yield of nutrients/kg$^{0.75}$ per day						
energy (kJ)	745	820	700	770	715	132
protein (g)	8.4	8.7	7.6	8.4	9.4	0.6

by Table 1.6 is that after scaling the species according to metabolic body size the daily yield of milk energy and protein from Friesian cow, Jersey cow, Saanen goat, sow and bitch are all very similar. The only species that appears atypical is *homo sapiens*. It must be said however that if the table had included, say, the Hereford cow and the mare they would have been intermediate between woman and the high yielders cited. The only point to be made at this stage is that the physiological capacity of the dairy cow to produce milk is no greater than that of a wide range of mammals.

The very low milk yield of women is, of course, linked to the very slow growth potential of the human baby. Those who like to speculate on these things may consider the nature of the link between our highly developed mental capacities as a species and our slow rate of physical development. Those who like to speculate on how to produce a more efficient cow by genetic means may consider what, if anything, has been achieved by selection in the modern Friesian cow when her capacity to produce milk solids is so like that of many other mammals. Of course, the biological efficiency of milk production is not the same thing as economic efficiency. Friesians can make money. However, in attempting to understand dairy cows in general and, in particular, whether one cow is better than another, we need to take these fundamental laws of biology into account.

Development of today's dairy cow

The domestication of primitive oxen is a fascinating topic but one reserved for historians. The development of the dairy cow is predominantly a twentieth-century phenomenon. Prior to the twentieth century, problems of hygiene, transport and inequalities of wealth made dairy farming a minority occupation; most cattle were bred for work, beef, leather, dung or status according to the ecology and mores of their owners. Refrigeration, rapid transport, a scientific understanding of fermentation processes, a greater concern for the welfare of the people and an awareness of the nutritive value of milk and milk products have all combined to encourage the development of the specialist dairy farm.

The visible outcome of this has been more milk from bigger cows in larger herds. This has been achieved by a combination of breeding, feeding and management. The relative importance of each will emerge as the book progresses. At first glance the importance of breeding and genetics is obvious, perhaps too obvious. Friesians give more milk than Jerseys and much more milk than Herefords. Milk from Channel Islands breeds, Jerseys and Guernseys, contains more fat than that from Friesians. Moreover, annual lactation yield in any one herd may range from 3000 to 1100 litres. Clearly, under similar conditions of feeding and management, cows produce milk that differs both in quantity and quality and these traits are heritable.

Genetic progress with dairy cattle is inevitably slow because of the long time taken first to demonstrate genetic superiority and then to breed from superior individuals. Woolliams has calculated the theoretical rates of genetic improvement in the performance of dairy cattle using conventional and improved progeny testing to be 1.5 to 2.0 and 2.6 to 3.3% per annum respectively rising to approximately 4% if progeny testing is accompanied by multiple ovulation with embryo transfer (MOET) (Hill, Thompson and Woolliams, 1990). These points will be discussed in detail in Chapter 11.

Actual increases in milk yields and butterfat concentration in UK herds recorded by the Milk Marketing Board (MMB) over the 25-year period prior to the imposition of milk quotas are given in Table 1.7.

Table 1.7   Average yields of milk and butterfat concentrations in 1957 and 1981

	Dairy Shorthorn	Jersey	Friesian
Lactation yield: 1957	3627	3276	4464
1981	4834	3814	5490
1981:1957	1.33	1.16	1.23
Butterfat (%): 1957	3.61	5.22	3.83
1981	3.86	5.42	4.05
1981:1957	1.07	1.04	1.06

Milk yields increased by 23% in Friesians and 16% in Jerseys during that 25-year period. These increases are phenotypic, i.e. they reflect improvements due to nutrition and management as well as to breeding. Indeed, MMB calculates that the annual rate of genetic improvement has been only about 0.3% per annum. It is difficult to establish what improvements were achieved in the late 19th and early 20th centuries because few complete records exist for this time. There are several reports dating back to 1810 of Dairy Shorthorns giving 32 quarts (36 litres) of milk per day and one of a cow, Magdalena, who gave 45 quarts (over 50 litres)! I see no reason to doubt these figures but think it highly unlikely that these yields could have been sustained for long since the biological capacity of the cow to produce milk would soon have been constrained by her owner's inability to provide food of sufficient *quality* and her own inability to consume food in sufficient *quantity* to meet her metabolic needs. *It has always been the case that the capacity of the mammary gland of the high-yielding cow to produce milk has exceeded the capacity of the cow upstream of that mammary gland to provide the nutrients necessary for milk synthesis.* This will have been true for Magdalena presented with the range of foodstuffs available in 1810; it is equally true for the modern Holstein cow stimulated to even greater yields by the administration of bovine growth hormone (bovine somatotropin or BST).

The increase in the total lactation yield of dairy cows has been achieved more by nutrition than by genetics. Individuals like Magdalena have always been present in the population. On a low quality diet such cows would not only fail to achieve their full genetic potential for milk production but also probably be more prone to problems such as infertility and metabolic disease. In other words, they would be unsuited to their environment. As the quality of food available to dairy cows improved, in particular through improved techniques for forage conservation and reduced relative costs of purchased concentrates, so the environment tended to favour individuals (and selection for individuals) with higher lactation potential. This is illustrated in Table 1.8 which compares inputs and outputs for a typical 1910 Dairy Shorthorn giving 10 litres/day and a 1980 Friesian/Holstein giving 30 litres/day. It is assumed that forage for the Shorthorn is hay (8.5 MJ ME/kg DM) and for the Friesian it is grass silage (10.5 MJ ME/kg DM). The concentrate intake of the Shorthorn is taken from a recommended ration of the time, namely 3 lb bran, 1 lb wheatmeal and $1\frac{1}{4}$ lb decorticated cotton cake.

Relative to the 1910 typical Dairy Shorthorn, the 1980 Friesian is producing three times as much milk. She is 18% heavier. Her forage intake is 17% less in terms of dry matter but almost identical in terms

Table 1.8   Typical daily inputs and outputs in winter for a Dairy Shorthorn cow in 1910 and a Friesian/Holstein in 1980

	Dairy Shorthorn 1910	Friesian/ Holstein 1980	1980:1910
Body weight (kg)	550	650	1.18
Milk yield (l/d)	10	30	3.0
Forage intake*			
dry matter (kg)	10.8	9	0.83
metabolizable energy (MJ)	92	95	1.03
crude protein (kg)	0.92	1.4	1.52
Concentrate intake			
dry matter (kg)	2.4	9	3.79
metabolizable energy (MJ)	26.5	120	4.53
crude protein (kg)	0.56	1.6	2.86
Total daily intake/kg metabolic size ($W^{0.75}$)			
dry matter (g)	117	140	1.20
metabolizable energy (kJ)	1049	1670	1.59
crude protein (g)	13	23	1.77
Output, ml milk/kg dry matter in food	0.75	1.66	2.21

* forage assumed to be hay in 1910 and silage in 1980

of ME. She has, however, been given 4.5 times as much ME and 2.9 times as much crude protein in the form of concentrate food. Using the concept of metabolic size to compare these two animals we find that the Friesian is eating 20% more dry matter, 59% more ME and 77% more crude protein. Her ability to convert food dry matter to milk is 2.2 times that of the Shorthorn, principally because her intake of nutrients in excess of maintenance is so much greater. The improvement in nutrient intake has made it possible to obtain responses to selection for higher milk yield. Table 1.8 also makes the point that one of the highly advantageous correlated responses to selection for higher milk yield is a higher appetite.

Welfare of today's dairy cow

The UK codes of recommendations for the welfare of livestock, drawn up by the Farm Animal Welfare Council (1983) recognise that the welfare of any animal is determined by its physical and mental state. Good welfare implies therefore both a sense of fitness and well-being. The welfare 'needs' of an animal whether on farm, in transit or at the point of slaughter may be defined by 'five freedoms'. These are:

(1) Freedom from hunger and thirst – achieved by readily accessible fresh water and a diet to maintain full health and vigour.
(2) Freedom from discomfort – achieved by appropriate shelter with a dry, restful lying area and temperature within an acceptable range of tolerance.
(3) Freedom from pain, injury and disease – achieved by prevention or rapid diagnosis and treatment.
(4) Freedom to express normal patterns of behaviour – achieved by the provision of room to move, things to do and the company of their own kind.
(5) Freedom from fear – achieved by conditioning animals to their surroundings and avoiding situations which cause stress.

The welfare of animals in any husbandry system, old or new, may be evaluated in the context of these five freedoms, recognising that no system is perfect and that conflicts may arise – for example, the more comfortable cow bed may be one that carries an increased risk of mastitis. Where a conflict of this sort occurs I believe that the right to good health and vigour should be paramount, for farm animals at least. It inevitably follows that I place less importance than some on freedom of behavioural expression. This is an unfashionable view among those concerned with the welfare of animals but one which is particularly necessary when evaluating the welfare of the dairy cow. Relative to many farm animals she is far less likely to suffer problems as a consequence of having nothing to do all day. On the contrary, she is worked quite extraordinarily hard. I have already indicated the fact that the dairy cow takes in a very high quantity of energy by consuming large amounts of rich food and that most ME is used to do work and then dissipated as heat (see Fig. 1.2). To match the daily work output of a Friesian cow giving 35 litres milk per day, every day of the week, a man would have to jog for 6 hours a day, every day of the week!

Improvements in the provision of nutrients and selection of cattle to meet these improvements has produced an animal that may be compared with a highly tuned racing car, designed to run as fast as possible on very high grade fuel. As with Grand Prix cars, the results are, at best, spectacular, but at least unreliable and at worst catastrophic. The old house cow may, by contrast, be compared with the family car – much slower, but much less likely to blow up. The farmer who feeds modern rich feeds to modern, highly selected dairy cattle has to achieve a fine balance between overfeeding and so provoking a digestive crisis, and underfeeding and so provoking a metabolic collapse. The term *production disease* has been coined to describe problems such as ketosis, laminitis and other painful and distressing conditions which have arisen as a direct consequence of the drive to produce as much milk as possible. These will all be considered in detail. For the moment, it is necessary only to point out that they contribute a major threat to her welfare.

Selection of cows for high milk production can lead to other correlated responses which may be deleterious to welfare. For example, selection for high yields inevitably means selecting for bigger udders. This problem has been exacerbated by the management decision to milk the high-yielding dairy cow twice or, at most, three times daily. The beef cow, with perhaps 20% of the yield of the dairy Holstein, suckles her calf five to seven times daily. Breeding, nutrition and management therefore all combine to produce pendulous udders which are chronically uncomfortable and carry an increased risk of teat injuries and mastitis. Even if not pendulous, bigger udders tend to distort the normal conformation of the hind legs and so predispose to problems such as foot lameness. It is no coincidence that over 70% of cases of damage to the soles of the feet of dairy cows occur in the lateral claws of the hind feet (Greenhalgh, McCallum and Weaver, 1981). It is important, therefore, to include conformation and other traits not directly associated with production in any selection index (Chapter 11).

Welfare considerations occur constantly throughout this book as an essential part of the understanding of the dairy cow. Many are problems of the moment which can be overcome by good stockmanship. The two problems I have cited here, poor conformation and production disease, are more fundamental, being unhappy side-effects of the anatomical and physiological traits actively encouraged by selection. If we are to consider the welfare needs of the animal as well as the efficiency of production in developing the cow of tomorrow, these consequences must never be overlooked.

Biological and economic efficiency

Table 1.6 showed that the physiological capacity of the dairy cow to produce milk is not greater than that of many other mammals; moreover that, relative to her size, the Friesian produces no more milk nutrients than a Jersey. Table 1.8 showed that, by the rather naive criterion of milk output:dry matter intake the 1980 Friesian is 2.2 times as efficient as the Dairy Shorthorn, due to a greater intake of nutrients relative to maintenance; this is largely a consequence of improvements in nutritive value revealing the genetic potential for high yields. It does not explain why the large, high-yielding Friesian and even larger, higher-yielding Holstein have come to dominate dairy production in areas of the developed, affluent world. The reason becomes clearer after inspection of Table 1.9, which examines further the case of the 1910 Dairy Shorthorn and the 1980 Friesian/Holstein whose biological efficiency was compared in Table 1.8.

The Friesian with a peak winter yield of 30 l/day is assumed to have a total lactation yield of 5800 litres. The Shorthorn's total yield of 2600 litres is relatively higher in relation to winter yield because it is assumed that she will produce relatively more milk off summer grass. For simplicity, however, milk price has been assumed to be the same (15p/1 in 1980). The rate of herd replacement is assumed to be 15% for 1910 Shorthorns and 25% for 1980 Friesians, which is reasonable since culling intensity has increased for a number of reasons which include genetics and disease. I have, however, set veterinary costs at £18 for both cows, there being no sound evidence to suggest that cows in 1910 were any healthier than now, or vice versa.

Table 1.9　Economic efficiency of a 1980 Friesian/Holstein and a 1910 Dairy Shorthorn compared at 1980 prices

	Friesian/Holstein (1980 model)	Dairy Shorthorn (1910 model)
Physical performance		
output (1/year)	5800	2600
grass and forage intake (tonnes DM)	4.6	5.1
concentrate intake (tonnes DM)	1.6	0.4
Economic performance (£/year)		
income: milk at 15p/l	870	390
calf value	80	60
less herd replacement*	75	34
gross income	875	416
Variable costs		
grass and forage	72	80
concentrate at £160/tonne	256	64
veterinary	18	18
miscellaneous	32	24
total	378	186
Gross profit margin	497	230
Fixed costs†	400	400
Net margin	+97	−170

* assumed 25% annual replacement of Friesians, 15% for Dairy Shorthorns
† includes labour, machinery, property and interest repayments

　　The modern dairy farmer spends about twice as much on variable costs, especially the purchase of compound foods (£378 *v.* £186 at current prices) to generate rather more than twice as much gross income (£875 *v* £416). In terms of gross profit margin (income minus variable costs) both cows are comfortably in profit. If the story stopped here then the modern dairy farmer could perhaps be accused of profiteering. However, intensive dairy farming in the developed world carries very high fixed costs for payment of labour, machinery, purchase and maintenance of buildings and repayment of loans with interest. The Milk Marketing Board estimated fixed costs in 1980 to be about £400 per cow. This figure, unfortunately, is approximately the same whatever the size and milk yield of the cow since she will require about as much building space and almost as much time and attention at milking. When fixed costs are a high proportion of total costs it almost invariably pays to produce as much milk as possible per cow. This means not only selecting the biggest, milkiest cows possible but also feeding them as much as possible even if it means buying in concentrate food at relatively high cost.

This general rule is open to some degree of manipulation and much of the remainder of this book is concerned with how this might be done in face of political and economic pressures such as the imposition of quotas or fluctuations in the relative prices of milk and concentrates. It would however be naive to overlook the enormous importance of fixed costs to the economic efficiency of modern dairy production although they will not be considered again until the final chapter. Equally, one cannot properly analyse the efficiency of milk production only in terms of net profit. The optimal utilisation of the resources of the land for the production of milk from dairy cows depends on a thorough analysis of all the major components of the system. These are:

(1) *The capacity of the land to provide nutrients.* This is largely a matter of climate, e.g. corn (maize) silage potentially yields more utilisable ME per hectare than grass silage but not in cool, wet areas like Ayrshire (Scotland) which is nevertheless magnificent dairy country.

(2) *The capacity of the cow to take in nutrients.* This depends on availability, quality and the digestive capacity of the cow. California, a land rich in corn and alfalfa, can support the Holstein; the dry, unirrigated tropics of Africa cannot. Here selection pressure favours the low input–low output animal.

(3) *The capacity of the cow to achieve and sustain high yields of milk solids.* Assuming a sufficiency of nutrients and good health, this is largely determined by genetics and physiology.

(4) *Lactating period as a proportion of total lifetime.* This depends upon age at first calving, subsequent fertility and age at slaughter.

(5) *Health and welfare.* Health problems such as infertility and mastitis obviously have detrimental effects on production. It is sometimes claimed that a cow cannot produce to peak capacity unless her

welfare is satisfactory but I have seen too many high-yielding cows with sore feet to be convinced by this argument.

(6) *Economics of production*. These are considered under sources of income, variable and fixed costs and the most consistent thing one can say about these is that they will continue to fluctuate. However, as indicated already, the biological laws that govern milk production by dairy cows are more consistent, predictable and innately sensible. By considering dairy production in terms of these fundamental laws I hope to provide a sound, logical framework that will enable the dairy producer to respond effectively to the whims and vaguaries of economic forces, or better still, to get one step ahead.

2 Digestion and Metabolism

Digestion is the process whereby food is broken down into substances which can be absorbed across the cell lining of the gut. The nutrients so digested and absorbed are then available to the tissues of the body for processes of metabolism such as work, growth and the synthesis of milk.

The dairy cow is a ruminant, which means, in the simplest terms, that prior to the stomach she has a very large fermentation vat, the reticulo-rumen, which contains microorganisms capable of breaking down the structural carbohydrates of plant cell walls into simple organic compounds that can be absorbed and used for energy metabolism.

The material that flows out of the rumen into the abomasum or true stomach contains food that has escaped fermentation and a rich source of protein from microorganisms which have multiplied within the rumen. From the abomasum downwards the digestive tract of the cow and the processes that take place therein are essentially similar to those in man and other simple-stomached animals (Fig. 2.1). It is the unique structure and function of the rumen that has given the dairy cow the capacity to subsist so largely on plant materials that are of no direct nutritional value to man.

Structure of the digestive tract

The cow is popularly assumed to have four stomachs, the reticulum, rumen, omasum and abomasum. In practice, it is easier to consider only two functional compartments. The first, which is made up of the reticulum and rumen, provides a huge reservoir wherein food, saliva and other secretions may be stirred and mixed in the presence of microorganisms. At birth the reticulo-rumen is smaller than the true stomach or abomasum which forms the second functional compartment. As the calf begins to consume dry food, the reticulo-rumen

UNDERSTANDING THE DAIRY COW

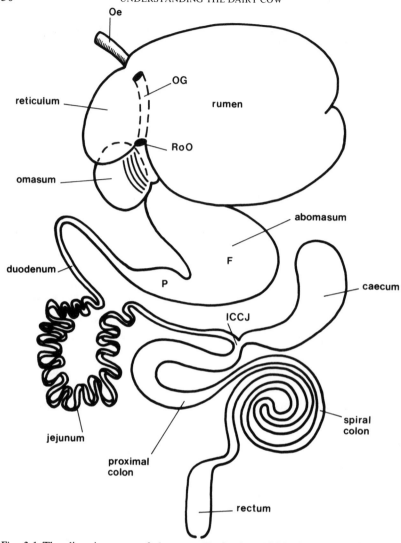

Fig. 2.1 The digestive tract of the cow. F=fundus, ICCJ=ileocaecolic junction, Oe=oesophagus, OG=oesophageal groove, P=pylorus, RoO=reticulo omasal orifice.

grows in volume and acquires a mixed population of microorganisms which are capable of breaking down the fibre in cell walls. In the adult cow it fills the major part of the abdominal cavity, extending on the left side from the diaphragm to the pelvic inlet (Fig. 2.2). The inner surface of the reticulum has a 'honeycomb' appearance. The inner wall of the rumen is made up of thousands of finger-like processes (papillae), the function of which is to increase the surface area for absorption of end products of ruminal fermentation (Fig. 2.3(a)). The oesophageal opening and the reticulo-omasal orifice, which form respectively the entrance and the exit to the rumen, are sited remarkably close to one another and linked by the oesophageal groove, the lips of which close when a young calf sucks milk, thereby creating a short cut from the oesophagus into the abomasum. The relationship of the rumen (or paunch) to the rest of the digestive tract is like a huge lake with a river running through one corner (see Fig. 2.1), a design particularly well suited to delaying the passage of food down the main stream of the gut and so allowing time for microbial fermentation.

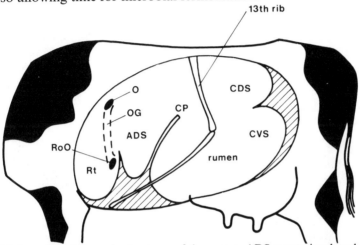

Fig. 2.2 Location of the main structures of the rumen. ADS = anterior dorsal sac, CDS = caudal dorsal sac, CVS = caudal ventral sac, CP = cranial pillar, O = oesophagus, OG = oesophageal groove, Rt = reticulum, RoO = reticulo omasal orifice.

The approximate relative weights and volumes of the major organs of the digestive tract of the adult cow are shown in Table 2.1. The weight of dried digesta (i.e. gut contents) removed from each portion of the gut is an indicator of volume. The percentage of total wet tissue

Table 2.1 Relative weights of portions of the gut and gut contents

	% total dry digesta in gut	% total wet tissue in gut wall
Reticulo-rumen	66	44
Omasum	4	16
Abomasum	5	7
Small intestine	15	20
Large intestine	10	13

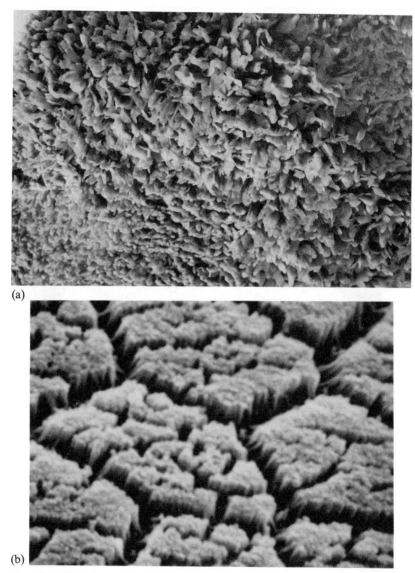

Fig. 2.3 (a) Visible papillae of the rumen wall. (b) Microscopic papillae of the small intestine wall (scanning electron micrograph).

in the gut wall describes the mass of each organ after removal of its contents.

The reticulo-rumen is by far the largest and heaviest organ in the gut. The 'third' stomach, or omasum, is also set aside from the mainstream of the gut which takes a short route between the reticulo-omasal orifice and the opening of the abomasum (see Fig. 2.1). The main body of the omasum is occupied, for the most part, by about 100 leaves of epithelial tissue so that while the mass of the empty omasum is relatively large (16%), it contains only about 4% of total digesta. This is packed between the leaves of epithelium and tends to be very dry, suggesting that the main role of the omasum is absorption of water and substances in solution. The exact function of the omasum remains something of an enigma. However, it can be removed surgically without any obvious effect on digestion and absorption which suggests that, as enigmas go, it is not very important.

The abomasum has two distinct sections: the fundus, which is the main site for the secretion of hydrochloric acid and enzymes that operate in an acid medium, and the pylorus, where material gathers before being propelled by muscular contraction through the pyloric sphincter and into the duodenum as a discrete bolus of partially digested food. The pyloric sphincter acts as a stopcock regulating the flow of material into the duodenum and small intestine.

The acid digesta are neutralised in the duodenum, subjected to further degradation by enzymes, then pass on to the jejunum and ileum which together make up the small intestine. The word 'small' is misleading. The volume of the contents of the small intestine is only about 15% of total digesta. However, the epithelial lining of the small intestine is made up of millions of microscopic finger-like processes called papillae giving it an enormous surface area relative to its mass (see Fig. 2.3(b)). Not only is the small intestine the major site for absorption of nutrients, it also contains some of the most metabolically active, hard-working cells in the body. It has been calculated that the daily rate of protein synthesis in the epithelial cells of the small intestine is equal to the total protein mass of the tissue. In other words, average life expectancy of these proteins is about one day. By contrast, the life expectancy of proteins in skeletal muscle is about one month.

The large intestine is made up of the caecum, another reservoir set aside from the mainstream of the gut, and the colon. Despite its relatively large volume, the contribution of the large intestine to digestion and absorption is small. It acts as a secondary site for microbial fermentation and for absorption of water and substances in solution such as mineral salts. The epithelium of the large intestine is not expanded by papillae to the same degree as the jejunum and ileum so its effective surface area is, in fact, less than that of the small intestine.

Eating

The grazing cow harvests grass with a mowing action. The tongue swings across the pasture pulling grass into the mouth where it is cut off by the action of the incisor teeth of the lower jaw against the dental pad (there being no incisor teeth in the upper jaw). This action makes the cow a much less selective grazer than the horse or even the sheep. The amount consumed at a mouthful obviously depends on the length of the grass. When the grass is relatively short and moist it can be swallowed almost at once with little chewing or ensalivation. The same applies to short-chopped silage. Long hay is pulled into the mouth and chewed between the molar teeth until it is reduced to a suitably small and moist bolus for swallowing. The upper jaw of the cow is wider than the lower, which means that only one side of the molar teeth is involved in chewing at any one time, which saves wear and tear. Cereals and pelleted feeds are consumed very quickly, usually without chewing. When cows are fed unrolled barley or corn it is common to see in the faeces grains that have been swallowed whole and passed unchanged right through the gut.

Since the teeth of cows are not well adapted to biting, they are not well equipped to deal with root crops. Cows have a tendency to swallow small tubers like potatoes whole, and occasionally choke themselves in the process. Large root crops such as turnips and fodder beet are often fed whole. When hungry dairy cows are competing with each other for a limited supply of (say) whole fodder beet (which they greatly relish), they do tend to swallow chunks of alarming size. I believe that it is both safer and fairer to slice root crops before feeding.

Salivation

The cow, in common with all ruminants, salivates copiously and continually. The rate of flow from all salivary glands is difficult to measure but is probably about 60 ml/min at rest increasing to 120 ml/min during eating and about 150 ml/min during rumination (Church, 1979). This means the amount of saliva secreted per day usually exceeds 140 l (over 30 gallons). The amount secreted *per minute* during eating and rumination is relatively independent of the dry matter content and physical form of the food. However, the time spent eating and ruminating a dry, long fibrous food like hay obviously greatly exceeds that spent eating the same amount of nutrients as cereals or pelleted concentrates. The total quantity of saliva secreted per day is thus very dependent on the physical form of the food consumed.

The saliva of ruminants contains large quantities of sodium bicarbonate and phosphate which act as buffers in the rumen to restrict the fall in pH (increase in acidity) that would otherwise accompany the production of organic acids as end products of rumen fermentation. When a cow eats food such as long hay, the time spent eating and ruminating is prolonged. This increases the flow of saliva, dilutes and buffers the contents of the rumen and so reduces the risk of acidosis.

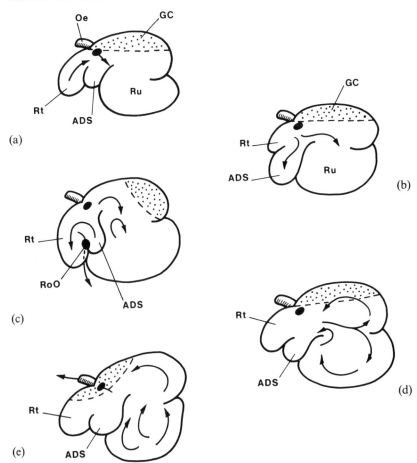

Fig. 2.4 Movements of the reticulo-rumen. ADS = anterior dorsal sac, GC = gas cap above rumen contents, Oe = oesophagus, Rt = reticulum, Ru = rumen, RoO = reticulo omasal orifice.
(a) Bolus arrives at ADS, Rt starts to contract. (b) Rt fully contracted. (c) Contraction of ADS, material forced through RoO. (d) Mixing contraction in main body of rumen. (e) Belching contraction.

Rumen movements

The movements of the reticulo-rumen are complex and elegant
because they govern four quite distinct processes:

(1) Mixing of rumen contents.
(2) Outflow through the reticulo-omasal orifice.
(3) Regurgitation of food for ruminating.
(4) Eructation of gases (belching).

Some of these movements are illustrated in Fig. 2.4 (a)–(e). The arrival
of a bolus of food stimulates a contraction of the reticulum which
forces the contents of the reticulum into the anterior dorsal sac of the
rumen (a). The liquid on top, and solids floating on that liquid, pass
over into the main body of the rumen, the rest of the liquid and the
denser solids remaining in the anterior dorsal sac (b). The anterior
dorsal sac then contracts as the reticulum relaxes (c), allowing some
material to return to the reticulum and some to pass out into the
omasum and abomasum through the reticulo-omasal orifice. The
wave of contraction then spreads to the main body of the rumen,
tending to stir the contents in a figure-of-eight motion (d). This basic
series of contractions designed to achieve mixing and outflow takes
place at intervals of about 40–60 seconds during eating and
rumination, slowing to about 80–100 seconds when no food material
is entering through the oesophagus. The time taken for food passing
over the cranial pillar to complete a cycle of the rumen is about 1 hour.
    Eructation is achieved by a reverse contraction which originates in
the caudo-ventral sac of the rumen and drives the gas cap on top of the
rumen contents up towards the opening of the oesophagus (e).
    Regurgitation of food material involves an isolated contraction of
the reticulum which forces the contents of the reticulo-rumen anterior
to the cranial pillar up against the gastric opening of the oesophagus
which is, at the same time, dilated by expanding the thorax, thus
permitting food and fluid to be drawn up into the mouth.
    The movements of the reticulum and anterior dorsal sac both
control the rate of outflow of rumen contents and sort out the solid
material into that which passes into the abomasum and that which is
returned to the body of the rumen. In essence, the smaller and denser
particles tend to pass out into the abomasum and the larger, lighter,
less digested particles are returned for further chewing and fermentation. The rate of passage of material out of the rumen depends
therefore not only on food intake but also on the composition of the

food. When a dry cow eats a maintenance ration consisting entirely of forage, about 3% of the rumen contents exit per hour through the reticulo-omasal orifice. Looked at another way, this means that undigested forage remains in the rumen for about 33 hours. When a high-yielding cow eats a mixture of concentrates and forage *ad libitum*, rates of passage for forage and concentrates respectively may be 6% and 10% per hour, equivalent to mean retention times of 16 and 10 hours.

Rumination

Rumination or 'chewing the cud' begins with the propulsion of reticulo-rumen contents into the mouth (Fig. 2.4 (e)). Most of the liquid is rapidly squeezed out and re-swallowed. The solids, which tend to be the longer, lighter, less digested material, are then chewed over at leisure. Cows tend to ruminate most when totally relaxed and have even been observed to do so when electroencephalographic (EEC) records show them to be in a state of deep sleep. The function of rumination is to work over food until it can escape the cycle of regurgitation and pass on into the abomasum for further digestion. The time spent ruminating is simply and closely related to the amount of chewing required to achieve this objective. A dry cow subsisting on a diet of long hay may ruminate for 8–10 hours per day. A beef animal being fattened on a diet made up entirely of cereals or a milled and pelleted compound diet may only ruminate for 30 minutes. Such animals may exhibit a form of 'sham' rumination, going through the motions of chewing without having regurgitated any food. Nobody really knows why they do this but it does not usually last very long and seldom, if ever, develops into a compulsive, stereotyped form of behaviour like bar chewing in confined sows or weaving in stabled horses. Veal calves denied all access to dry food do spend abnormally long periods of time in purposeless oral activity such as licking and chewing their wooden crates but this appears to be related more to the fact that they have nothing to eat rather than nothing to ruminate. Rumination therefore appears to be directed strictly to the pragmatic business of breaking down food. There is no reason to suppose that cows have an innate physiological need to ruminate and so would feel emotionally deprived by the absence of long fibre. However, rumina-tion of long fibre is a vital part of normal ruminant digestion because it stimulates the flow of saliva and so dilutes the rumen contents: it accelerates outflow into the abomasum and it neutralises the acids produced by fermentation. It is common knowledge among stockmen that dairy cows need long fibre. It is becoming quite common practice to incorporate sodium bicarbonate into high cereal, low fibre diets, sometimes in amounts as high as 400 g/day, to avoid problems of ruminal acidosis. When it is understood that a dairy cow ruminating 8–10 hours/day transfers about 2000 g/d of sodium bicarbonate into the rumen via the saliva, moreover that the sodium is returned to the body and the bicarbonate has come from a waste product, carbon dioxide, then it can be appreciated that it makes more sense to manipulate the diet so that the cow produces sufficient sodium bicarbonate for herself at zero cost.

Digestion

The digestibility of any food, or the nutrients contained within that food, is determined both by the physical and chemical form of the food and the digestive processes to which it is exposed. The measurement of *apparent digestibility* of, for example, organic matter (OM) is straight-forward since

$$\text{apparent digestibility of OM} = \frac{(\text{OM in food} - \text{OM in faeces})}{\text{OM in food}}$$

This simple equation forms the basis for most routine digestibility trials to evaluate diets for all animals including man in terms of their capacity to supply digestible energy and protein. In the typical digestibility trial to evaluate a diet for ruminants, four sheep are fed fixed, weighed amounts of the test diet for four weeks. After an 18-day adaptation period, faeces (and any food refusals) are collected over a 10-day period and stored under refrigeration. Analysis of food and faeces generates apparent digestibility values for the dry matter (DM) or OM in the diet or for specific components of the diet, e.g. energy, protein, fat or fibre.

Routine digestibility trials of this type are enormously useful but they have their limitations. In all species they fail to recognise the fact that faeces contain not only undigested food but also endogenous material secreted into the lumen of the gut. *True digestibility* defines the proportion of any nutrient that is actually digested and absorbed across the gut wall, thus,

true digestibility of OM
$$= \frac{OM \text{ in food} - (\text{faecal OM} - \text{endogenous faecal OM})}{OM \text{ in food}}$$

It follows that true digestibility always exceeds apparent digestibility. Endogenous losses are difficult to measure or estimate with precision, not often attempted in practice and outwith the scope of this chapter (but see McDonald, Edwards and Greenhalgh, 1988).

In ruminants, measurements of apparent digestibility also fail to distinguish between microbial digestion, principally in the rumen, and enzymic digestion in the abomasum and duodenum. Rumen microbes make energy and protein available to their host, the cow, by fermenting that which they can (e.g. carbohydrate, including fibre but not fat) and incorporating organic nitrogen into microbial protein which is washed out of the rumen for digestion downstream. Before embarking on a detailed examination of how the processes of digestion convert food for animals into substrates that can be absorbed and used for energy and protein metabolism, it is, I think, helpful to distinguish four feed fractions. These are:

(1) Material which is quickly fermented or degraded in the rumen. Material (carbohydrate and protein) which is fermented to energy-yielding substrate such as volatile fatty acids is termed *quickly fermented energy* (QFE). Organic nitrogenous material that is rapidly degraded to ammonia is *quickly degradable nitrogen* (QDN).
(2) Material which is slowly and thus incompletely fermented or degraded in the rumen is termed SFE or SDN.
(3) Material which is unfermented or undegraded in the rumen is termed unfermentable, digestible energy (UDE) or undegradable, digestible nitrogen (UDN).
(4) The final fraction is that which is neither fermented nor digested. Since it does not contribute to energy or protein supply it requires no code.

These terms will be explained in more detail *en route* and used as the basis for Part II, *Feeding the Dairy Cow*.

Fermentation of carbohydrates in the rumen

Fermentation was defined by Pasteur as 'life without oxygen'. In essence, it describes the anaerobic metabolism of energy sources by micro-organisms to provide fuel for the work they have to do, which is mainly synthesis of protein for growth and multiplication. This involves the addition of a high energy phosphate bond to adenosine diphosphate (ADP) to create adenosine triphosphate (ATP), the energy currency for all biological systems, microorganisms, plants or animals.

For practical purposes carbohydrates are the only nutrients that can be fermented in the absence of oxygen to yield significant amounts of energy as ATP to the microbes. All that needs to be known of the chemistry of carbohydrates is that they are made up of simple monosaccharides with 6 carbon atoms called hexoses (e.g. glucose, $C_6H_{12}O_6$) or 5 carbon atoms called pentoses (e.g. ribose, $C_5H_{10}O_5$). Disaccharides are made up of two monosaccharides, e.g. sucrose (commonly called sugar) is glucose + fructose; lactose, the disaccharide in milk, is glucose + galactose. Larger, more complex polymers are called polysaccharides; the most important of these are starch, pectin, cellulose, hemicellulose and lignin. They occur both inside plant cells and in the cell walls.

Sugars and starches in the cell contents ferment very rapidly in the rumen. A very small amount of energy is captured by the microbes as ATP, most is excreted by them in the form of simple organic acids, acetic, propionic and butyric, containing respectively 2, 3 and 4 carbon atoms, which serve as the principal source of dietary energy to the host animal, namely the cow. In normal, neutral conditions (pH = 6 – 7) these volatile fatty acids (VFAs) exist in the rumen as their sodium salts, the sodium arising from saliva and direct secretion of body fluids across the rumen wall (Fig. 2.5). Effectively all the VFAs produced as an end product of microbial fermentation are absorbed directly across the rumen wall, which explains the need for the large surface area created by the rumen papillae. Indeed the production of VFAs especially acetate, is essential to the normal development of the rumen papillae and so proper absorption.

The rate of fermentation of all sugars and most starches is so rapid that they are completely broken down to VFAs. However, a few carbohydrate-rich foods such as maize and tapioca contain starches that ferment more slowly and therefore incompletely in the time available to them in the rumen, some material passing unchanged

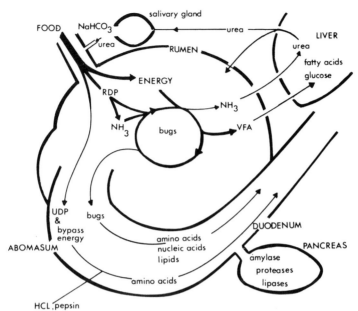

Fig. 2.5 Digestion of energy and protein in the rumen and abomasum. RDP = rumen degradable protein, UDP = undegradable dietary protein, VFA = volatile fatty acids.

through the reticulo-omasal orifice and into the abomasum where it can be broken down by acid digestion to monosaccharides like glucose. Of course, if the plant cell walls are not ruptured by mechanical action or microbial digestion then the carbohydrates inside the cell are protected both from fermentation and subsequent acid digestion, which is why it is not uncommon to see whole barley or maize grains in cattle faeces.

Plant cells walls contain the following polysaccharides in decreasing order of digestibility: pectin, hemicellulose, cellulose and lignin. Lignin, which is, in effect, wood, is completely resistant to fermentation or acid digestion. The rate of fermentation of the other carbohydrates depends on their accessibility to microorganisms in the complex structure of the cell wall. Pectins ferment rapidly and may be included with starches and sugars under the heading of *quickly fermentable energy* (Table 2.2). Hemicellulose and cellulose constitute *slowly fermentable energy*. In general their rate of fermentation depends on the degree of lignification of the cell wall. The extent of fermentation of this fraction is determined therefore both by microbial fermentation rate and by retention time in the rumen. Hemicellulose and cellulose that pass out of the rumen cannot be broken down by acid digestion although they will, of course, meet another population

Table 2.2   Categories of nutrients

Organic matter in food	Carbohydrate			Protein	Fat
	cell contents	cell walls			
	starch, sugar	pectin, hemicellulose, cellulose	lignin		
Rumen fermentation	Quickly fermented	Slowly fermented	none	partial degradation	none
Acid digestion	digestible (if escape fermentation)	indigestible		digestible	digestible
Feed Analysis	N-free extract		Crude fibre	Crude protein (N × 6.25)	Oil (ether extract)
	Acid-detergent solubles		Acid-detergent fibre		
	ND solubles	Neutral detergent fibre			

of microorganisms in the caecum and colon where further fermentation can take place.

The digestion of protein will be considered later. As far as the microorganisms are concerned, dietary protein is a source of nitrogen for microbial protein synthesis but yields very little energy as ATP. Fats cannot be metabolised to produce ATP for microorganisms in the absence of oxygen. Indeed, excessive amounts of fats in ruminant diets can inhibit fermentation, slow down digestion and so reduce appetite unless they are covered in a protective coat that resists attack by microorganisms.

Farmers wishing to know the nutritive value of the forage they produce or the cake they buy must rely on measurements of chemical composition made by agricultural chemists. Ideally this would include all elements of the carbohydrate fraction as illustrated in Table 2.2. In practice this would be too expensive, so commercial agricultural chemists opt for simpler forms of analysis.

The practical evaluation of feeds for ruminants will be considered in detail in Chapter 6. For the moment it is sufficient to say that the conventional procedure for proximate analysis divides the carbohydrate fraction into *crude fibre* (CF), which describes the residue left after digestion of the sample in strong alkali then strong acid, and *nitrogen-free extract* (NFE), which is the remainder after all other determinations have been made. This procedure has given good service in the past but is now out-of-date since CF and NFE do not correspond precisely to the carbohydrate chemistry of the food nor to the physiological description in terms of quickly fermentable, slowly fermentable and unfermentable energy (Table 2.2). Van Soest (1982) in the USA has pioneered more elegant techniques for the description of carbohydrates based on their solubility in detergents.

Neutral detergent fibre (NDF) describes total cell wall carbohydrate, with the exception of pectin. In Table 2.2, quickly fermented energy corresponds to carbohydrates that are soluble in neutral detergent. Slowly fermented energy equals NDF-lignin.

Currently in the UK, the official procedure for describing fibre is a modification of Van Soest's acid-detergent fibre method. This modification, affectionately known as MAD fibre, produces a residue intermediate between NDF and CF. It may be a little more useful in practice than CF but has little theoretical meaning since it too does not correspond either to specific carbohydrate chemistry nor to processes of fermentation in the rumen.

The complete sequence of reactions involved in the fermentation of carbohydrates is of no great practical concern. The final stages do however have an important bearing on our understanding of food utilisation and of milk quality.

In effect, for practical purposes, all carbohydrates ferment via monosaccharides to the 3-carbon compound, pyruvic acid. In the simplest case this may be written

$$C_6H_{12}O_6 \longrightarrow 2CH_3COCOOH + 4H \ (+ 2ADP \longrightarrow ATP)$$
glucose $\qquad$ pyruvic acid

One mole of glucose has been converted to 2 moles of pyruvic acid. This has stored a small amount of energy for the microbes as 2 moles of ADP have been converted to ATP. It has also accumulated four protons or hydrogen ions (4H). Pyruvic acid and protons are both very unstable and are instantly metabolised further to a mixture of acetic, propionic and butyric acids plus the stable end products carbon dioxide, water and methane.

Acetate production, $CH_3COCOOH + H_2O \rightarrow CH_3COOH + CO_2$
$+ 2H \ (+ 4\frac{1}{4}ATP)$

Propionate production, $CH_3COCOOH + 4H \rightarrow CH_3CH_2COOH$
$+ H_2O \ (+ 4ATP)$

Butyrate production, $2CH_3COCOOH \rightarrow CH_3CH_2COOH + CO_2$
$(+ 3ATP)$

Methane production, $CO_2 + 4H_2 \rightarrow CH_4 + 2H_2O$

Acetate production generates a further excess of protons (H); propionate production, on the other hand, requires protons. Methane

(CH_4) production 'mops up' excess protons according to the balance between H production from fermentation to pyruvate and H uptake from fermentation to propionate. The yield of ATP to the micro-organisms from fermentation of glucose to acetate, propionate and butyrate is approximately $4\frac{1}{4}$, 4 and 3 moles ATP/mole glucose, i.e. so far as the microorganisms are concerned fermentation to acetate is the most efficient process.

So, why is this important?

Table 2.3 Fate of 1000 kJ energy fermented in the rumen

	Fermentable energy in diet (kJ)		
	Hay 1000 Cereal 0	Hay 500 Cereal 500	Hay 200 Cereal 800
Energy yield to cow (kJ)			
acetate	448	395	322
propionate	150	260	393
butyrate	182	185	185
total VFA	780	840	900
Energy yield to microbes (kJ)	70	60	50
Energy loss as methane (kJ)	150	100	50

Table 2.3 illustrates what happens to 1000 kJ of energy fermented in the rumen of a cow given different mixtures of forage (which is mainly slowly fermented energy from cell walls), and cereal (which is mainly quickly fermented energy from starch). In each case the vast majority of fermentable energy is unavailable to the microorganisms and is absorbed by the cow as VFA (780–900 J/kJ fermented energy). The amount of energy actually used by the microorganisms for their own purposes is very small (50–70 J/kJ). The amount of fermentable energy converted to acetate or propionate is highly dependent on the balance between quickly and slowly fermented carbohydrate in the ration. When hay is the only food the energy yield as acetate is about three times that as propionate but 150 J/kJ fermented energy are lost as methane. This is a substantial loss. A dairy cow eating food with an average price of £100 per tonne dry matter belches about £46 worth of methane per year. Given a population of 4 million dairy cows and heifers in the UK that represents £184 million worth of hot air! Increasing the ratio of cereal to forage increases the ratio of propionate

to acetate and reduces energy loss as methane. The ratio of propionate to acetate can also be manipulated by the use of selective antibiotics to manipulate the microbial population of the rumen. Monensin, for example, has been extremely effective in increasing the efficiency of feed utilisation by beef cattle by increasing propionate:acetate and reducing energy losses as methane.

In the dairy cow, however, it is nearly always essential to feed a diet that includes sufficient structural, slowly fermented carbohydrate to ensure a high yield of acetate relative to propionate despite the associated high loss of energy as methane. Firstly, acetate is the main precursor of milk fat. When cows get insufficient slowly fermentable carbohydrate (digestible fibre), e.g. when they are fed excess cereals or young spring grass, the fat content of milk falls because acetate production is reduced. Secondly, fermentation patterns that generate a high ratio of acetate to propionate are most efficient at providing energy for the microbes (see Table 2.2) and so providing microbial protein for the cow. Thirdly, when fermentation of carbohydrates to VFA proceeds at a steady, controlled rate and is accompanied by a steady flow of buffers in saliva, the rumen liquor is weakly acidic but stable at a pH above 6. The full range of rumen microbes can thrive in these conditions. The more microbes the more effective is fermentation, the more complete is the breakdown of digestible fibre and the better the appetite of the cow becomes. When a cow eats a large cereal meal, starches are fermented so rapidly that the rumen liquor becomes increasingly acid (pH may fall below 5). This kills off large numbers of normal rumen microbes and allows organisms like lactobacilli, which are favoured by low pH, to predominate. This upset to the normal rumen population will, at the very least, reduce appetite. At worst it can lead to severe metabolic disorders or even sudden death (Chapter 7).

To summarise, most of the digestion of carbohydrates in ruminants involves microbial fermentation to VFAs and most of this takes place in the rumen. A few starches that escape fermentation can be broken down to monosaccharides by enzymes in the abomasum or released from the pancreas into the duodenum (see Fig. 2.5). A small proportion of the fibre that escapes fermentation in the rumen may be fermented to VFA in the caecum and colon. However, this usually accounts for less than 10% of energy supply.

Digestion of crude protein

This specific role of proteins in food is to provide amino acids as building blocks for protein synthesis in the animal that eats that food. Only some, such as lysine, threonine, histidine, methionine and cystine are essential; others can be manufactured in the tissues of the body from these essential amino acids.

The main source of essential amino acids to the cow is the population of microorganisms that leave the reticulo-rumen and are digested in the acid medium of the abomasum by pepsin or in the neutral medium of the duodenum by pancreatic proteases (see Fig. 2.5). Synthesis of protein by microorganisms in the rumen does not, however, necessarily require amino acids from true protein as building blocks; most bugs, in fact, preferentially use simpler nitrogenous compounds like ammonia (NH_3).

The concentration of nitrogenous compounds in foods for ruminants is expressed as *crude protein*. This value is derived simply by measuring the nitrogen (N) concentration in dry matter and multiplying it by 6.25 on the assumption that dietary protein contains 16% N (1/6.25). This assumption is only approximate. Moreover the term crude protein does not distinguish between true protein and non-protein nitrogen (NPN) sources like urea which can make a significant contribution to the needs of the microbes for protein synthesis. It also does not distinguish nitrogenous compounds that are bound up in the lignified and therefore totally indigestible portion of the cell wall and so are unavailable for degradation in the rumen or subsequent acid digestion. This fraction is best described by acid-detergent insoluble nitrogen (ADIN) (Van Soest, 1982).

To understand protein digestion in the ruminant it is necessary to recognise these three components of dietary crude protein: true protein, NPN and ADIN (Fig. 2.6). The flow of dietary crude protein to the rumen is augmented by recycled NPN in the form of urea secreted from the blood into the digestive tract via saliva and also across the rumen wall (see Fig. 2.5).

Micro-organisms attack nitrogenous compounds entering the rumen, degrading them more or less successfully to simple amino acids and NH_3 before incorporating them into microbial protein. In normal circumstances at least 70% of microbial protein is synthesised from ammonia. There have been a few heroic experiments to demonstrate that cows can produce milk, at least for a while, on diets containing no true protein at all and therefore no amino acids. These are quoted as

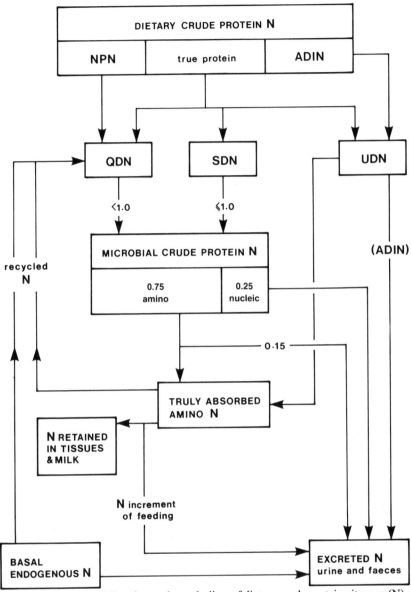

Fig. 2.6 Pathways of digestion and metabolism of dietary crude protein nitrogen (N). ADIN = acid-detergent insoluble N, NPN = non-protein N.

further evidence, if further evidence were needed, that cows do not compete with man for the same sources of food. However, they overstate the case. Rumen microbes undoubtedly require a small supply of amino acids as essential building blocks for protein synthesis. When the diet contains only NPN they obtain these amino acids by degrading endogenous protein sloughed from the walls of the rumen. This cannot, of course, be maintained indefinitely.

The rate at which rumen microbes degrade incoming crude protein depends on its chemistry and availability. NPN and water soluble true proteins degrade completely almost instantaneously to NH_3. This is *quickly degradable nitrogen* (QDN, Fig. 2.6) and becomes analogous to quickly fermented energy as defined earlier. The remainder of the true protein fraction degrades more slowly and is therefore *slowly degradable nitrogen* (SDN). As with slowly fermented energy, the extent of degradation depends upon the *rate* of degradation by rumen microorganisms and on retention time in the rumen. In other words, as food intake increases and rate of passage of food through the rumen increases, so the SDN fraction tends to decrease. The fraction that escapes degradation in the rumen, *undegradable dietary nitrogen* (UDN), consists of true protein which can be digested in the abomasum and duodenum, and ADIN which is completely indigestible.

The rate of synthesis of microbial crude protein depends on two things: the supply of degradable nitrogen and the supply of fermentable energy.

Microbial protein synthesis

The rate of synthesis of microbial protein is linked closely to the rate of fermentation of organic matter (especially carbohydrate) in the rumen.

The maximal theoretical yield of microbial organic matter is approximately 26 g/mole ATP
Microbes contain about 50% protein, itself containing 16% N
The maximum ATP yield from fermented organic matter (FOM = 25 moles/kg
Thus the maximum yield of microbial protein N is:

$$26 \times 0.5 \times 0.16 = 2.0 \text{ g/mole ATP}$$
$$\text{or} \quad 2 \times 25 \quad\quad = 50 \text{ g/kg FOM}$$

The cow does not have access to all the microbial protein synthesised in the rumen, only that fraction which passes out into the abomasum. However, all the VFAs produced by fermentation are absorbed directly across the rumen wall and are thus available to the cow as metabolizable energy (ME). If there was no outflow from the rumen to

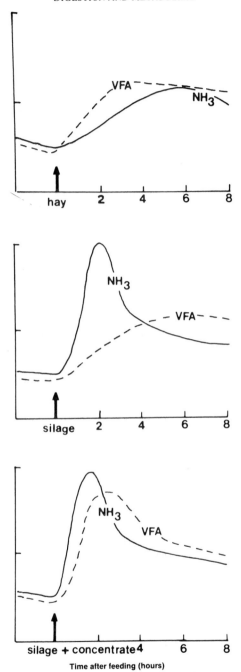

Fig. 2.7 Relative rates of production of ammonia (NH_3) and volatile fatty acids in the rumen of dairy cows after feeding. The units in the vertical axis are arbitrary and intended only to illustrate the balance between fermentation and protein degradation.

the abomasum the supply of microbial crude protein to the cow would be zero. The absolute supply of microbial crude protein, and the ratio of protein to energy supply, are determined by the ratio of microbial protein outflow from the rumen to total protein synthesis. Various factors affect this, of which the most important is the rate at which liquid contents, containing microorganisms, leave the rumen. For a cow at maintenance with a rumen outflow rate of 3% per hour the yield of microbial protein N may be 20 g/kg FOM, or 40% of total microbial protein synthesis. A high-yielding cow eating to appetite may have an outflow rate of 8% per hour and a protein yield of 40 g/kg FOM. These figures are approximate and likely to be modified slightly in the light of new research. They are however crucial to the proper protein nutrition of dairy cows, as will be explained in due course.

The relative rates at which energy and protein sources are respectively fermented and degraded depend on the relative proportions of the quickly and slowly degradable fractions. This is illustrated by Fig. 2.7 which shows what happens to concentrations of NH_3 and VFA in the rumen following meals of hay, grass silage or grass silage plus a small amount of quickly fermentable energy (QFE), e.g. fodder beet. The meal of hay alone produces a slow, relatively small increase in NH_3 and VFA concentrations. The similar size and shape of the two curves indicate that fermentation and protein degradation are relatively slow and almost in balance. In these circumstances all the degraded N may be incorporated into microbial protein, i.e. the ratio of degradable N to microbial N is 1.0.

When the cow eats silage only, fermentation and degradation get out of balance since silage contains large amounts of QDN but no sugars and almost no other sources of QFE. Not all the ammonia produced by degradation of QDN can be taken up by the microbes. This ammonia crosses the rumen wall, enters the veins leaving the rumen and reaches the liver where it is converted into urea (see Fig. 2.5). Circulating urea can return to the rumen across the wall and via saliva so ammonia produced in excess of the immediate requirements of the rumen microorganisms is not necessarily lost but can be recycled for use later on. It has been suggested that provided the total daily supply of degradable N does not exceed the capacity of the microorganisms to ferment energy sources (i.e. not more than 20–40 g degradable N/kg FOM), then all degradable N can be incorporated into microbial protein at an efficiency of 1.0 (Agricultural Research Council, ARC 1980, 1984). This has to be an exaggeration. When QDN greatly exceeds QFE, as in the case of the all-silage diet, blood urea concentration increases. Since the kidney excretes urea in direct proportion to blood urea concentration, some dietary QDN must be lost. For practical purposes we may conclude that when degradable nitrogen is not present in excess of FOM the efficiency of conversion of SDN to microbial protein N is 1.0 but the efficiency of conversion of QDN is less than 1.0. ARC (1980) assumed an efficiency of 0.8 for NPN. Since bugs see this merely as one form of QDN we may, for the time being, take the efficiency of incorporation of QDN to be 0.8 (see Fig. 2.6). Once again this may alter slightly in the light of new knowledge. When a silage diet is reinforced by a little QFE energy and nitrogen digestion in the rumen is better balanced (Fig. 2.7) thereby ensuring more efficient digestion of both classes of nutrients.

Urea recycling

When rumen microbes degrade urea to ammonia and then incorporate the nitrogen into protein, they do not discriminate between urea of dietary origin and endogenous urea recycled from the bloodstream to the rumen via the saliva and across the rumen wall (Figs. 2.5, 2.6). This endogenous urea may arise from:

(1) excess NH_3 absorbed earlier from the rumen and converted to urea in the liver;
(2) urea arising from the breakdown of body protein, especially during periods of under-nutrition;
(3) urea produced during the conversion of truly absorbed amino N to meat or milk protein at an efficiency below 1.0 (i.e. the N increment of feeding in Figure 2.6).

Consider the case of a cow in the dry season in the tropics with nothing to eat but mature standing hay which is very high in fibre and low in nitrogen. Such food will sustain neither the maintenance requirements of the cow, nor those of the rumen microbes. The cow breaks down some of her own protein reserves in (for example) muscle and skin to provide energy. The oxidation of protein produces urea which is not all lost (as it would be in simple stomached animals like ourselves) but recycled to the rumen, reincorporated into microbial protein, digested and reabsorbed as essential and non-essential amino acids. The importance of this physiological recycling of urea is threefold:

(1) it restores amino acids to a malnourished animal;

(2) it restricts urea excretion and therefore water loss (Chapter 4);
(3) it maintains the microbial population of the rumen.

If the microbial population of the rumen were to fall through lack of N its capacity to ferment energy would decrease, fibrous food would remain undigested in the rumen for longer periods and appetite would fall. Loss of appetite in an animal already barely able to extract sufficient digestible nutrients for maintenance would be catastrophic. In a strictly evolutionary sense, therefore, the most important of the consequences of the physiological urea cycle is this ability to use end products of protein metabolism to sustain the rumen microbial population and so sustain appetite for poor quality food. However, the highly productive dairy cow can also derive a significant proportion of her QDN for microbial protein synthesis from urea produced during the synthesis of milk protein from absorbed amino acids.

Truly absorbed amino nitrogen

Microbes and undegraded dietary nitrogen (UDN) are digested in the abomasum and duodenum. Microbial crude protein consists of about 75% amino acids and 25% nucleic acids. The latter cannot be used by the tissues of the body for protein synthesis. The true digestibility of microbial crude protein is taken as 0.85.

Thus the supply of *truly absorbed amino nitrogen* (TAAN) to the cow from microbial protein N is (see Fig. 2.6):

0.75 × 0.85 microbial protein N = 0.64

UDN contains two fractions. ADIN is almost completely indigestible. The remainder appears to have a true digestibility of 90%. Thus digestible UDN (DUN) becomes:

$$DUN = 0.9 (UDN - ADIN) \tag{2.1}$$

This is assumed to consist entirely of amino acids. Thus,

$$TAAN = 0.64 \text{ microbial protein N} + 0.9 (UDN - ADIN) \tag{2.2}$$

So far, this says nothing about the *quality* of the amino acid supply. The quality, or biological value, of a dietary protein is determined by its composition of amino acids relative to the amino acid requirements of the animal that eats it. Microbial crude protein has a high biological value for growth and lactation in cattle. The quality of UDN of course reflects the original food and so can vary greatly (Chapter 6).

TAAN defines the amount of dietary N that is digested and absorbed by the cow and should not be confused with apparently digestible N (or digestible crude protein, DCP) which simply describes the difference between N intake and N loss in faeces.

Throughout this section I have used N rather than CP to signify organic nitrogenous compounds which can be used in protein metabolism and this is undoubtedly the better term since it permits clear distinctions, e.g. between NPN, protein N and ADIN. However, nutritionists have been used for so long to thinking and working in units of crude protein that in Part II, *Feeding the Dairy Cow*, I shall have to do the same. Hereafter I shall call the crude protein equivalent of TAAN *metabolizable protein* (MP). Similarly, the terms QDN, SDN and DUN will become QDP, SDP and DUP.

Digestion of lipids

The lipids that we eat ourselves are mainly fats and oils. Fats, which are mostly of animal origin, are esters of glycerol and fatty acids in which all the carbon bonds are saturated with hydrogen atoms. The structure of stearic acid, $C_{17}H_{35}COOH$, may be written:

$$CH_3.CH_{2(16)}.COOH$$

Oils are unsaturated, which means that one or more carbon atoms lack H atoms and are linked by double bonds. Thus linoleic acid, $C_{17}H_{31}COOH$, which has two double bonds, may be written:

$$CH_3.CH_{2(14)}.CH_{(2)}.COOH$$

Vegetable oils, from soya beans, sunflower seeds, etc., are made up of unsaturated fatty acids which exist in liquid form at room temperatures. In cold-blooded animals like fish, fatty acids are mostly unsaturated, otherwise they (the fish) would seize up. Fats and oils serve mainly as energy sources, although all mammals have a small specific requirement for polyunsaturated fatty acids such as linoleic.

The metabolism of fats and oils to yield energy as ATP can only take place in the presence of oxygen. It cannot occur therefore in the anaerobic environment of the rumen. Moreover, since fermentation of carbohydrates generates an excess of protons, unsaturated oils entering the rumen tend to pick up excess protons and so leave as saturated fats. This is why fats in cow's milk and in the meat of beef and lamb are highly saturated. I shall deal briefly with the problem of saturated and unsaturated fats in human nutrition in Chapter 13.

So far as the cow is concerned, she can digest fats post-ruminally using lipases from the duodenum and pancreas. She can satisfy her needs for essential fatty acids by digestion of lipids in microbial cell walls. A small amount of dietary fat (up to about 5%) can pass through the rumen, becoming saturated in the process but without effect on normal fermentation. If it is deemed expedient to give fat in excess of this, e.g. to boost milk supply, then it must be protected from microbial action – and they must be protected from it – by coating the fat in some undegradable material. I must here re-emphasise the point that protected fats, although an excellent source of ME, cannot be fermented and do not therefore contribute to the energy supply for microbial protein synthesis. Thus some attempts to increase milk yield by adding fat to increase ME supply may have failed because they reduced the supply of TAAN or metabolizable protein.

Absorption and secretion of minerals

The major minerals, calcium (Ca), phosphorus (P), magnesium (Mg) and sodium (Na), and the minor elements such as iron, copper, cobalt, etc., may be fed in inorganic salts or organic complexes. Some digestive processes may be involved in releasing minerals from organic complexes but the main factor determining the *net* uptake of minerals from the gut to the body is the balance between secretion into the gut and true absorption from it. This is illustrated in Fig. 2.8 for sodium. The exact figures are, of course, subject to considerable variation according to circumstances. The important message in Fig. 2.8 is the *relative* flow of Na between the compartments of the body. In this example the cow eats 40 g Na per day. The amount of Na entering the rumen through saliva and across the rumen wall is 500 g/day. Thus only 8% (40/500) of Na entering the gut is of dietary origin. 540 g/day Na passes out of the rumen and a further 60 g enters the duodenum and jejunum. From here on net movement becomes outward, 580 g/day passes out of the terminal ileum and large intestine, leaving only 20 g to exit in the faeces. It should now be clear that it is quite inadequate to describe Na uptake from the gut by saying that apparent digestibility, or, more usually in the case of minerals, *apparent availability*, is 50% [(40–20)/40].

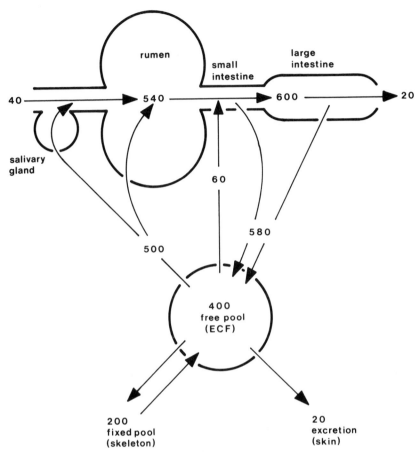

Fig. 2.8 Secretion and absorption of sodium (g/day) between the digestive tract and extra-cellular fluid (ECF) of a dry cow.

About two-thirds of the sodium in the body is in the extra-cellular fluid (ECF), indeed the Na ion is mainly responsible for maintaining the osmotic pressure of this freely-exchangeable fluid pool. The total Na contained in the ECF is, in this example 400 g. The remainder is bound up in bone and can be released only slowly. This compartment may be called the *fixed pool*.

The amount of Na leaving the small intestine per day (600 g) is 50% greater than that of the freely exchangeable Na in the body. This highly dynamic state can be maintained in perfect equilibrium unless something such as diarrhoea occurs to upset reabsorption of Na from the terminal ileum and large intestine. The scouring cow or calf rapidly becomes dehydrated not only because she loses water; failure to reabsorb Na from the gut reduces the osmotic pressure and thereby the volume of the ECF. If water and Na are not replaced by rehydration therapy the animal will suffer peripheral circulatory failure (which is a precise definition of 'shock') and so die.

Calcium exchange

Over 99% of body calcium (Ca) is contained in the mineralised skeleton (bones). Only about 12 g is in the ECF or freely exchangeable pool. Figure 2.9, keeping things as simple as possible, examines the daily exchanges of Ca (a) in a dry cow, (b) in early lactation. The dry cow eats 30 g Ca/day and excretes 29 g/day in the faeces. 5 g/day of Ca are absorbed from the gut and 4 g/day returned to it. This balance between absorption and excretion is under control of the parathyroid hormone and vitamin D_3 [1,25 2(0H) cholecalciferol] which also acts as a hormone. When the need for Ca is high these hormones increase absorption and decrease secretion across the gut wall. When the need is low, as in the case of the dry cow, absorption is low and most of the Ca is returned to the gut. In this cow, in exact Ca balance, 1 g/day is excreted in the urine.

At the onset of lactation, this balance is greatly disturbed because suddenly the cow starts to secrete large quantities of Ca into the milk. In Fig. 2.9(b) the cow is secreting 30 g Ca/day in 25 litres/day of milk. Her intake of food in general has increased, increasing Ca intake to 65 g/day. PTH and vitamin D_3 are now very active, increasing the net movement of Ca out of the gut to 27 g/day. They have also brought about a much smaller net utilisation of 2 g/day Ca from bone. This cannot, of course, persist indefinitely. The amount of Ca in the freely exchangeable ECF pool is 12 g. If this is not replaced, milk secretion could, in theory, drain the ECF of Ca in 10 hours!

Everybody who knows cows knows of the problem of 'milk fever'; collapse and paralysis following calving leading rapidly to death unless the cow is treated with large quantities of calcium. Knowing the dynamic state of Ca in the body helps us to understand why. Chapter 7 will consider practical problems associated with acute and chronic deficiencies of the major minerals, calcium, phosphorus and magnesium, and with the trace elements. There is little that one needs to know of the general principles of physiology of digestion and metabolism of the trace elements save that they are, on the whole, stored in the body in very large amounts relative to daily dietary intake and slowly

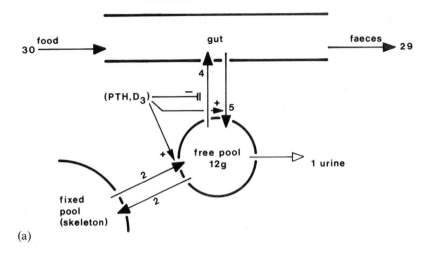

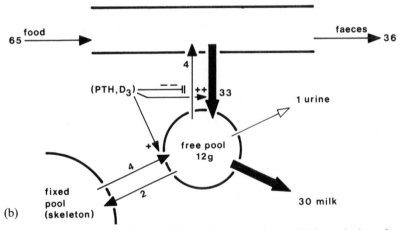

Fig. 2.9 Exchanges of calcium (g/day), (a) in a dry cow, (b) in early lactation. PTH = parathyroid hormone, D_3 = vitamin D_3 or cholecalciferol.

excreted. Unlike Na, Ca and Mg therefore, problems of deficiency tend to be slow in onset but problems of toxicity can occur if high intakes are maintained for long periods.

Energy metabolism

Energy metabolism is the process whereby an animal first oxidises nutrients absorbed from the gut (metabolizable energy, ME) to generate the energy currency of the body, chiefly by converting ADP to ATP, and then uses ATP to do work. All the work done by the body, using ATP as a fuel, including visible tasks like walking into the cowshed and invisible tasks like milk protein synthesis in the mammary gland or active secretion of enzymes into the gut, generates heat. Moreover, about two-thirds of the energy of absorbed nutrients is dissipated as heat during the processes of intermediary metabolism that ultimately consume oxygen, produce carbon dioxide and water and so generate ATP. In other words, the efficiency of conversion of ME to ATP is only about 30–40%. The amount of heat produced during synthesis of ATP from any given chemical substrate is independent of the routes of intermediary metabolism. We may therefore describe energy metabolism in the lactating cow very simply:

$$ME = H + RE_l \pm RE_g$$

where H is heat production,
RE_l is the amount of energy secreted in milk (l is used to indicate lactation because this argument is going to contain quite enough Ms as it is), and
RE_g refers to energy gained in or lost from body tissues (protein and fat).

At risk of stating the obvious, I should add that once ME has been converted to heat it is lost to the system. It cannot be recycled, which is why requirement for energy is so high relative to other nutrients.

Efficiency of utilisation of ME

It is not easy to distinguish experimentally between the energy costs of synthesis of ATP from ME and the energy expended as ATP does work. We can however do something rather similar and rather more practical which is to conduct an *energy balance trial*. A set of animals are fed known amounts of ME, measurements are made of H (or RE) and the unmeasured quantity obtained by simple subtraction. For cattle, it is measured using an animal calorimeter which either records heat loss directly or estimates it from oxygen consumption and carbon dioxide and methane production (Blaxter, 1967). Direct measurement of RE requires the technique of comparative slaughter which involves killing animals at different stages of an experiment and doing chemical

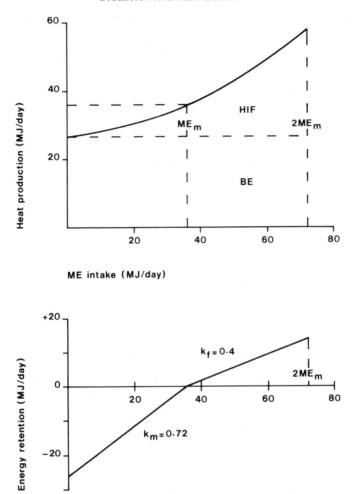

Fig. 2.10 Effect of increasing ME intake on energy balance in a 300 kg Friesian heifer.

analyses on all or part of the body to measure changes in energy content. This approach is, to say the least, prohibitively expensive. It also suffers from the fundamental problem that an animal can only be killed once. There is much current research interest in techniques for estimating body composition (and thus *RE*) in the living animal (Lister, 1984) but no real prospect yet of a technique applicable to cattle that compares in precision with calorimetry.

Let us consider what happens when we feed increasing amounts of ME to a 300 kg Friesian heifer (Fig. 2.10). When intake is zero, or more specifically for a ruminant, during the 3rd–4th day without food (Blaxter, 1967), heat production is 26 MJ/day This is defined as *fasting metabolism* or *basal metabolic rate* (BE). Increasing ME intake inevitably increases H. This is called the *heat increment of feeding* and may be expressed in absolute units (MJ) or as a proportion of ME (MJ H/MJ ME).

When ME intake = 36 MJ/day, ME = H. The heifer is now in exact energy balance. Her *maintenance requirement* for ME is therefore 36 MJ/day. HIF below maintenance has been 10 MJ or 0.28 MJ/MJ ME. As ME increases above maintenance (which we may call ME_m for short) so H increases at an accelerating rate. An increase in ME from 36 to 72 MJ/day, i.e. from maintenance to twice maintenance, or ME_m to $2ME_m$, increases H from 36 to 57.6.

Thus:

$$\text{HIF } (ME_m \rightarrow 2ME_m) = 57.6 - 36 = 21.6 \text{ MJ}$$

The relationship between ME and H is curvilinear; H increases at an increasing rate with increasing ME

In this example, when ME = $2ME_m$ = 76 MJ/day,

$$H = 57.6 = 26(BE) + 31.6(HIF)$$

Let us now turn the equation round and examine the effect of increasing ME intake on RE, i.e. measure the efficiency with which increments of ME promote energy retention in the body. Let us further simplify matters by describing the curvilinear relationship between ME and RE in terms of two straight lines:

(i)   below maintenance ($BE \rightarrow ME_m$)
(ii) above maintenance ($ME_m \rightarrow 2ME_m$)

Above maintenance, the *net efficiency* with which ME is used to provide energy retention in the body is 14.4/36 = 0.40. This efficiency term is usually given the symbol k_f to denote efficiency of fat deposition. For practical purposes we may assume that k_f is unaffected by the ratio of protein to fat in body gains so remains constant throughout growth.

Many new feeding systems for ruminants define food quality in terms of *net energy for gain* (NEg), which is the capacity of the food to promote energy retention/kg when fed above maintenance. The old Starch Equivalent system was the same in principle, although it did not use energy units but related all foods to NEg for 1 lb of starch. It follows that:

$$\text{NE requirement for gain} = k_f ME \text{ (above } ME_m)$$

The net efficiency of utilisation of ME below maintenance is given the symbol k_m. The meaning of this term is a little more difficult to grasp. In essence, k_m describes the efficiency with which ME substitutes for body tissues in providing ATP to do the work of maintenance.

It should also follow that:

NE requirement for maintenance = basal metabolic rate
or $BE = k_m ME_m$

In the simple case of the growing heifer, the relationship between ME intake and energy requirement for maintenance and gain may be written:

$$RE_f = k_f\left[ME-\left(\frac{BE}{k_m}\right)\right]$$ (2.3)

or, more simply still:

$$RE_f = k_f (ME-ME_m)$$ (2.4)

To calculate the capacity of ME to promote weight gain we need to know the amount of energy retained per unit of gain, EV_g (MJ/kg). This depends on the relative proportions of fat (39 kJ/g), protein (23.5 kJ/g), ash and water, both of zero energy content. Thus,

Liveweight gain (kg/day) = RE_g/EV_g

In the next most simple case, i.e. a cow giving milk but not pregnant and neither gaining nor losing energy

$$RE_l = k_l(ME-ME_m)$$ (2.5)

where k_l is net efficiency of utilisation of ME, *fed above maintenance* for milk synthesis.

The energy value of milk (EV_l) is determined by concentrations of the energy containing organic compounds lactose, protein and fat per kg milk.
Thus:

Yield (kg/day) = RE_l/EV_l

In real life the efficiency terms k_m, k_f and k_l vary according to diet quality. Moreover, in most lactating cows we need to take account of the fact that RE_g is unlikely to be zero and will include, in due course, the energy costs of pregnancy. For the moment, however, let us not complicate matters further.

Protein metabolism

Even a partial description of the nature and function of the individual body proteins is way beyond the scope of this book. I shall, rather arbitrarily, distinguish only four general types:

(1) *Structural proteins* in muscle; these include striated muscle or 'meat' and smooth muscle coating gut, uterus, arteries and other visceral organs.
(2) *Globular proteins*; mostly the enzymes which regulate the individual chemical reaction of metabolism.
(3) *Immune proteins*; immunoglobulins which recognise, and attempt to protect the tissues of the body from, invasion by foreign material such as harmful microorganisms.
(4) *Exportable proteins*, e.g. milk, wool.

Protein metabolism involves two distinct processes:

(1) Synthesis and degradation of proteins from their constituent amino acids. With globular proteins such as enzymes this takes place at great speed. The life expectancy of a protein in liver tissue is about one day, in striated muscle about one month. Free amino acids released by degradation of proteins can be recycled into new protein but the process of protein turnover (synthesis and degradation) is energetically expensive (about 5 kJ/g).
(2) Oxidation of amino acids to CO_2, H_2O and urea, yielding energy as ATP. In simple-stomached animals like man, urea is the main irreversible end point of the oxidation of protein N and is excreted in the urine. However, in ruminants, urea re-enters the gut in large quantities where it can be incorporated into new microbial protein, digested in the abomasum and reabsorbed into the blood as new amino acids (see Fig. 2.6). In the ruminant therefore, almost all protein N of dietary or endogenous origin is potentially available for recycling.

Efficiency of utilisation of MP

The efficiency with which metabolizable crude protein (MP) (or to be more precise, truly absorbed amino N) can be used to meet the needs of the body for protein synthesis depends on:

(1) MP supply in relation to requirement.
(2) 'quality', i.e. amino acid composition of MP.

(3) endogenous losses of crude protein (CP).

If a cow were fed a diet containing, say, 500 g CP/kg DM she would obviously absorb amino acids at a rate greatly in excess of her needs. A very large proportion of MP would be used as an energy source. Urea produced by oxidation of amino acids would not be reincorporated into microbial protein because they too would be faced by an excess of QDN so the majority of amino acids initially absorbed from the gut wall would be broken down and lost. In these circumstances the efficiency of incorporation of MP into body protein would be very low. The net efficiency of conversion of MP into body protein (k_n) is only maximum when MP supply is equal to, or less than, protein requirement, which is itself defined by the physiological need and capacity of the animal for protein deposition.

There are several ways of describing protein quality of foods (McDonald, Edwards and Greenhalgh, 1988). All methods, in essence, measure the efficiency of conversion of dietary protein into body protein during growth when intake is less than requirement. This depends on the balance of amino acids in the food relative to the overall balance of amino acids required by the animal for all aspects of protein synthesis. The individual who consumes protein of the highest possible biological value is the cannibal, provided he does not restrict himself fastidiously to muscle but consumes the whole carcass. Failing cannibalism, whole fish meal is a near perfect protein for growth, and for the same reason. The main milk protein, casein, is also excellent for growth, indeed it is usually adopted as the standard against which other proteins may be compared, and is, by definition, the perfect protein for milk synthesis.

The loss of endogenous protein is more difficult to grasp. Classical nutritionists, working with their black and white boxes, saw no problem because they chose not to look. They simply measured faecal protein N in the fasting animal, or estimated it by extrapolation, and called it metabolic faecal N. To this they added fasting urinary N and called that basal endogenous N loss (BEN), or in terms of crude protein, basal protein metabolism (BP). Both BP and fasting heat production (BE) can be criticised on biological grounds because, in an attempt to keep things simple, the animal has been subjected to the most unstable conditions of a prolonged fast to obtain a number that is then applied to a highly productive state like lactation. To describe BP simply in terms of fasting N loss in urine and faeces fails totally to account for the fact that rumen microbes do not discriminate between degradable dietary N, recycled urea of dietary origin, and endogenous urea arising from oxidation of body proteins. Indeed, unless degradable N of dietary origin is present in the rumen in great excess relative to fermentable organic matter, the microbes are certain to recycle endogenous urea.

If this were not so, we could describe the efficiency of utilisation of MP using an equation similar to those for ME (2.3, 2.4):

$$R_p \leqslant k_p(\text{MP}-\text{BP}) \tag{2.6}$$

Here the sign $\leqslant$ indicates that k_p for any stated dietary source of MP is only maximum when MP intake does not exceed requirement. In life, MP, BP and k_p, thus defined, will all vary according to the extent of incorporation of endogenous N into microbial protein, which makes it impossible for nutritionists to put any fixed value on MP or k_p for a particular diet. For example, increasing incorporation of endogenous N into microbial protein reduces BP, increases the ratio of MP to dietary crude protein, but markedly reduces the apparent efficiency term k_p.

Ørskov (1981) and his colleagues have achieved a neat experimental solution to this circular argument. They entirely sustain cattle by infusing into the rumen and abomasum chemicals corresponding to the end products of digestion (VFA, casein, minerals, vitamins, etc.) This creates an animal with no microbial fermentation system and indeed no functioning digestive system at all. In these circumstances,

$$\text{BP} = 2.2 \text{ g crude protein/kg bodyweight}^{0.75} \text{ per day}$$

BP is related to metabolic body size ($W^{0.75}$) whether within or between species in the same way as BE or ME_m. In other words, the value 2.2 BP/kg $W^{0.75}$ per day can be used, as a first approximation, to calculate endogenous protein loss for any weight of mammal *assuming no recycling*.

Substituting a figure of 2.2 g BP/kg $W^{0.75}$ in equation 2.6 gets round the problem of recycling by taking it out of the equation altogether. In these idealised circumstances,

(1) MP is assumed to be entirely of dietary origin.
(2) k_p describes the efficiency of use of MP as a function only of (a) diet quality, (b) specific requirement (e.g. growth or lactation).

What these assumptions lack in physiological elegance they make up for in practicability, since they make it possible to describe the protein quality of foods in a way that is largely independent of the physiological state of the animal.

This new concept of basal endogenous protein loss creates a further insight. It is difficult to measure protein synthesis (as distinct from deposition) with precision but direct and indirect estimates confirm that this too is a reasonably constant function of metabolic body size for mammals. (I have written more on this elsewhere, see Girardier and Stock, 1983.)

Protein synthesis = 15 g/kg $W^{0.75}$ per day
Since basal protein loss = 2.2 g/kg $W^{0.75}$ per day,
Efficiency of turnover of body protein = (15–2.2)/15 = 0.85

The efficiency of utilisation of dietary amino acids for protein deposition (k_p) cannot exceed 0.85. How close it gets to this figure depends on how well the supply of essential amino acids meets the overall amino acid needs of the body for the synthesis of specific proteins in specific tissues.

Regulation of food intake

So far this chapter has been concerned with what happens to food when it enters the body of the cow. To complete the story we need to consider what motivates the cow to eat (and stop eating). A vast array of individual factors has been implicated in the control of appetite, such as blood concentrations of glucose, VFA, individual amino acids, fatty acids, peptide hormones, rumen distension, liver temperature, body weight – the list is almost endless (Rook and Thomas, 1983). For our immediate purpose all we need to say is that food intake, *is* regulated and that three distinct factors are involved:

(1) The metabolic needs of the tissues for nutrients ('hunger').
(2) The size and digestive capacity of the rumen ('rumen fill').
(3) The hedonistic effect of highly palatable food ('appetite').

Hunger

A pregnant or heavily lactating cow will normally eat more food than a dry, barren one because she is doing more work. Nutrients are oxidised faster, causing a more rapid rate of decline in the concentration of the circulating metabolites that directly or indirectly tell the brain that the tissues are hungry. A healthy cow adapts to an increase in physiological demand for nutrients by an increase in *hunger* for nutrients which she will tend to eat, within reasonable limits, irrespective of taste.

Rumen fill

There is an obvious physical limit to the amount of food an animal can hold in its gut at any one time. This is particularly relevant for ruminants which eat a high proportion of slowly digestible and indigestible fibre which may remain in the rumen for many hours. Since the volume of the rumen is fixed (at least in the short term), intake is constrained by the rate at which material leaves the rumen either across the rumen wall or through the reticulo-omasal orifice. To use, once again, an unlikely example to illustrate a general truth: suppose a cow yielding 30 l milk/day is given only hay to eat. While this yield persists the metabolic needs of the tissues will greatly exceed the flow of nutrients from the gut, even though the cow consumes hay to the limits of rumen fill. She is in the unfortunate position of being full up but still hungry. *For most high-yielding dairy cows fed a substantial proportion of forage or other slowly digestible fibre, gut fill is the major constraint on nutrient intake and therefore on milk production.* As lactation proceeds and milk yield declines, so it becomes progressively easier for a cow to meet her metabolic needs within the constraint imposed by rumen volume. Whenever gut fill is the first constraint on food intake, the capacity of a cow to ingest nutrients is directly related to the digestibility of the diet, since any increase in digestibility:

(1) increases the *concentration* of metabolizable nutrients in the diet.
(2) is associated with a reduction in the concentration of indigestible and slowly digested fibre and so reduces the average *time* that food remains in the rumen.

Appetite

All animals from time to time eat more food than is strictly necessary. In these circumstances their food intake is not driven by hunger for essential dietary elements but *appetite* for foods that are pleasant to eat.

Appetite, thus defined, depends on the *palatability* (or taste) of the food on offer. Appetite can be manipulated by other factors such as novelty and competition. Cows given a midday feed of sugar beet pulp tend to eat more than those given a third feed of the same dairy cake that they get morning and evening in the milking parlour. The sight of another cow eating is a further spur to appetite.

The three determinants of food intake – metabolic need or hunger, rumen fill and appetite – are not entirely independent. For instance, it is possible to stimulate appetite by improving palatability or introducing the novelty of a third feed even when the main constraint on intake is rumen fill. In this case an increased appetite for a food of constant digestibility can only be accommodated by increasing rumen volume. In effect the cow decides that the satisfaction achieved by increasing food intake more than compensates for any dissatisfaction produced by increased distension of the gut. One of the arts of feeding high-yielding dairy cows is to persuade them to eat as much as possible. To achieve this, we need to understand not only her digestive physiology but also something of her psychology.

3 Reproduction and Lactation

The dairy cow is an exploited mother. To exploit her we have to ensure that she grows up, becomes pregnant, gives birth and lactates. All these processes are controlled by an elegant hierarchy of nerves and hormones. Once again, it is necessary for our purposes only to investigate these control mechanisms in sufficient depth to understand what is normal and how these normal mechanisms can be manipulated in practice.

Anatomy of the female reproductive tract

The gross anatomy of the reproductive tract of the cow, and its relationships to other body structures, are illustrated in Fig. 3.1. The vagina lies over the pubic bone which forms the ventral (lower) border of the pelvic canal. The main body of the uterus appears when examined from the outside *post mortem* or palpated via the rectum to be about 15 cm long, but internally is only 3–4 cm in length. The uterus is largely made up of two horns. The inner surface of the uterus is studded with about one hundred cotyledons, protuberances having a shape like leather buttons about 15mm in diameter in the non-pregnant animal. These form the sites for attachment of the foetal membranes of the developing calf during pregnancy. The oviducts, or Fallopian tubes, are over 20 cm long and coil back towards the bones of the pelvic canal. The ovaries are rugby-ball shaped and about 4 × 2.5 cm in size in the non-pregnant cow.

Follicle development and ovulation

Fertilisation, the fundamental act of reproduction, is achieved when a sperm carrying the male gamete and containing the male genes penetrates the female egg or ovum, fuses with the female gamete and

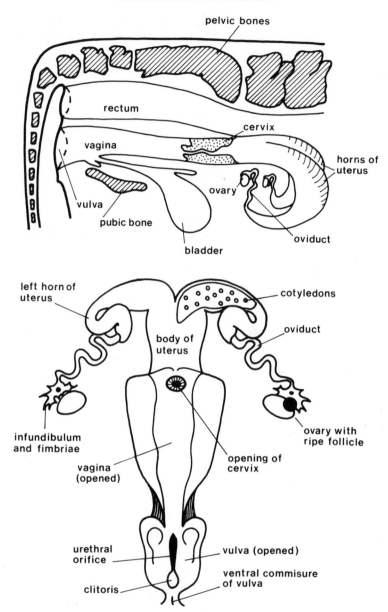

Fig. 3.1 The reproductive organs of the cow.

so lays down the blueprint for the creation of a new individual. What happens in both sexes before fertilisation is simply directed towards bringing male and female gametes together. What happens afterwards in the female is simply directed to maintaining this new life. Parturition (in this case, calving) may be the most momentous aspect of reproduction for the mother. So far as the life of the offspring is concerned, it represents little more than a change of environment and a change from intravenous nutrition to feeding by mouth.

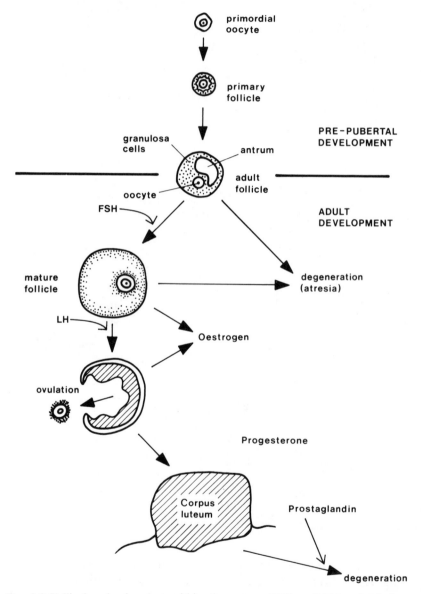

Fig. 3.2 Follicular development within the ovary. FSH = follicle stimulating hormone, LH = luteinising hormone.

The genetic material contained within every ovum released by a cow throughout her life is formed and stored within the ovary very early in development, indeed long before she herself has been born. These primordial oocytes, present at birth, acquire a coating of specialised granulosa cells and develop during prepubertal life into adult follicles (Fig. 3.2) with the oocyte situated towards one side and the centre of the follicle (the antrum) filled with follicular fluid. After puberty, gonadotrophic hormones from the anterior pituitary act to mature the follicles in sequence, usually one at a time and so establish the rhythm of the oestrous cycle in the mature, non-pregnant female. Initially, *follicle stimulating hormone* (FSH) causes one follicle to enlarge and mature so that, just prior to ovulation, it is a turgid, fluid-filled structure distending the wall of the ovary and about 20 mm in diameter. As the follicle enlarges the granulosa cells secrete increasing amounts of *oestrogen*, the gonadal (or sex) hormone that induces the

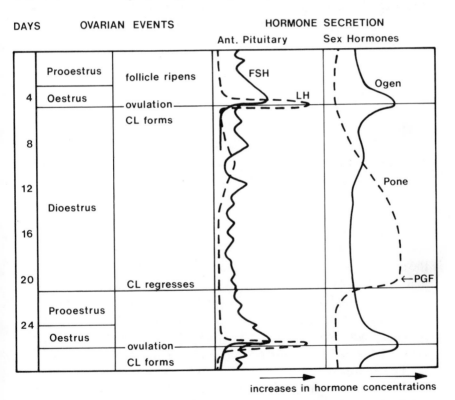

Fig. 3.3 The oestrous cycle of the cow. CL = corpus luteum, FSH = follicle stimulating hormone, LH = luteinising hormone, Ogen = oestrogen, Pone = progesterone, PGF = prostaglandin $F_2\alpha$.

physiological and behavioural changes associated with being in oestrus, on heat or 'bulling'. Actual rupture of the ripe follicle and release of the ovum is induced by a sudden large release of *luteinising hormone* (LH) from the anterior pituitary gland (Figs. 3.2, 3.3). This also controls the reorganisation of the ruptured follicle into a new structure, the *corpus luteum*, or yellow body, with a new function, the secretion of the hormone *progesterone*.

The number of ova that are normally released during the adult life of a cow is far less than the number of primordial oocytes present in the ovary at the time of birth. Normally the majority of follicles do not develop to the point of ovulation but degenerate (atresia, Fig. 3.2). There is no obvious explanation for this quirk of evolutionary development. It does mean, however, that with appropriate hormonal control the cow can be induced to release far more ova than she would normally achieve in a lifetime. Induced multiple ovulations with embryo transfer (MOET) has become a recognised technique in commercial cattle breeding. It will be reviewed, somewhat critically, in Chapter 13.

The oestrous cycle

Oestrus describes the time when the female cow is sexually attractive and attracted to the male and will therefore permit mating. The duration of oestrus is about 10–30 hours depending on factors such as breed, season and body condition and unless the cow becomes pregnant it recurs at intervals of about 21 (16–24) days. Since oestrus is the most obvious stage of the cycle to the observer, it is usual to consider this as Day 1. Figure 3.3 begins its description of the cycle about 3 days earlier, at about the beginning of *pro-oestrus*. FSH from the anterior pituitary stimulates (usually) one follicle to mature. The maturing follicle releases increasing amounts of oestrogen. After about two days the bull is able to sense by smell and other sensations that the cow will be on heat. She will not accept him at this time and he will probably not press his suit too strongly. There is however good evidence that he keeps a mental diary of cows that will be worth a visit on the morrow.

The cow in oestrus will normally show a wide range of sexual activities and responses. Usually, in a dairy herd, these do not involve the bull because he isn't there. She is restless and may make sexual overtures to other cows. More crucially, she will stand to be mounted by other cows. Practical recognition of oestrus from behavioural and other signs is covered in Chapter 12.

Ovulation, which is induced by a sudden large release of LH, usually occurs just after the end of 'standing heat' which is why it is better to be slightly too late than too early when using artificial insemination. The granulosa cells of the ruptured follicle immediately reduce their secretion of oestrogen (although oestrogen secretion doesn't stop, Fig. 3.3). As the corpus luteum develops it secretes increasing amounts of progesterone, the hormone primarily responsible for the maintenance of pregnancy. The luteal phase of the cycle, or dioestrus, lasts on average 16 days (12–20). At the end of this time the corpus luteum regresses and stops secreting progesterone. Current understanding is that it doesn't just die but is 'murdered' by another hormone *prostaglandin* (to be pedantic, prostaglandin $F_2\alpha$, or $PGF_2\alpha$) which is naturally secreted by the lining of the non-pregnant uterus and carried directly to the ovary through a counter-current structure in the local blood supply. In any event, administration of $PGF_2\alpha$ as a drug has proved a very successful practical technique for terminating dioestrus and bringing cows on heat within about three days.

Puberty

Puberty in the heifer occurs when she comes on heat and ovulates for the first time. Increased activity from the pituitary gland prior to puberty, in particular pulses of secretion of LH, stimulate some follicles to luteinise and secrete progesterone even though ovulation has not occurred. The 'feedback' on the pituitary from progesterone appears to potentiate LH (and FSH) release to the point where they can induce sufficient follicular development to achieve ovulation. Once triggered, oestrous cycles will then normally continue until the heifer is fertilised and becomes pregnant.

Factors affecting the onset of puberty are age, body weight and body condition. Consider these (rather dated) figures from 1963.

(1) Age and weight at puberty in dairy breeds

	Friesian	Breed Jersey	Shorthorn
Age at puberty (days), mean	400	360	335
range	290–595	235–555	190–475
Weight at puberty (kg), mean	250	165	240
range	200–340	120–225	215–295

(2) Effect of nutrition on age and weight at puberty in Holstein heifers

	Plane of nutrition High	Low
Age at puberty (days)	440	710
Weight at puberty (kg)	530	580

In the early 1960s, Friesian heifers tended to ovulate for the first time at about 13 months of age and about 250 kg body weight, Jerseys younger and Shorthorns younger still. The range of ages at puberty was much greater than the range of weights. Increasing plane of nutrition had a major effect on age at puberty but little effect on weight. The modern Friesian heifer probably reaches puberty at an age of about 11 months but still at a weight close to 250 kg.

These figures indicate that age, *per se*, has a relatively minor direct influence on puberty; it is far more important that the heifer should attain a size *and a degree of body fatness* that will enable her to sustain a pregnancy. Indeed, in cattle and many other mammals, the onset of puberty correlates more closely with body fat content than anything else. I would add that, as things stand at present, the growth of heifers to puberty should be normal and uncomplicated by steroids or other anabolic agents. Attempts to accelerate growth, puberty and first calving in heifers by the use of anabolic steroids have been catastrophic: lots of caesarian sections and practically no milk.

Fertilisation

The ovum or oocyte released from the Graafian follicle is directed by the fimbriae into the oviduct (see Fig. 3.1). Passage down the oviduct normally takes about 90 hours and proceeds in a 'to and fro' fashion, as muscle contractions move fluids in the oviduct alternately towards and away from the uterus. Sperm are usually deposited in the anterior vagina and outer portion of the cervix during natural service. The artificial inseminator usually penetrates the cervix and deposits sperm within the body of the uterus. The first sperm often arrive in the upper oviduct within 2–3 hours of mating. Since, with natural service, this usually occurs several hours before ovulation, the most successful sperm are in place and ready for the ovum before it arrives in the oviduct. This is aided by the movement of the cilia in the walls of the oviduct. Prior to ovulation they beat towards the ovary helping to move sperm upwards. After ovulation they beat towards the uterus.

'To and fro' contractions of muscle in the oviduct probably help to cluster large numbers of sperm and the oocyte into the same region. Sperm undergo a maturation process (capacitation) and are attracted towards the cells and membrane (zona pellucida) surrounding the ovum. A large number of sperm attack the zona pellucida mechanically and chemically using the enzyme hyaluronidase. Finally, perhaps 12 hours after ovulation and 24 hours after insemination, one sperm penetrates the zona and so brings together the male and female gametes. As soon as this first penetration has occurred the zona pellucida almost instantaneously changes in character so as to block the entry of any more sperm, one of the most remarkable features in this entirely remarkable process.

Pregnancy

The fertilised oocyte begins its process of cell division and differentiation as it migrates from the oviduct to the uterus. The process of differentiation establishes the cells of the embryo itself and the membranes that surround it, protect it and attach in to the uterus of its mother. At this stage the developing embryo is called a *blastocyst*. Figure 3.4 illustrates the early stage of formation on the two membranes and fluid-filled cavities ('bags of water') that surround the developing calf throughout pregnancy. The inner, *amniotic cavity* will eventually surround the calf in a relatively thick, lubricant, shock-absorbing fluid formed from the calf's urine and other secretions. A yolk sac, containing nutrients to feed the embryo, is present in early development but rapidly regresses (Fig. 3.4 (a), (b)). The second cavity is created by enlargement of the *allantois* which also acts as a reservoir for the excretion products of the developing calf and is the first 'bag of waters' to rupture at the time of calving. Very early in development the embryo has to signal to its mother its presence in the uterus to prevent

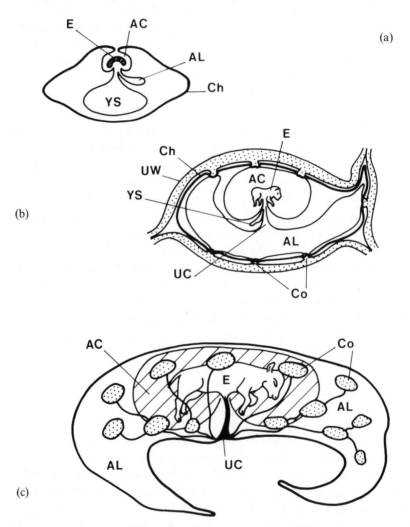

Fig. 3.4 Development of the calf, placenta and foetal membranes. (a) Blastocyst pre-implantation at 20 d. (b) Implantation complete at 40 d. (c) Foetal membranes at 100 d. AC = amniotic cavity, AL = allantoic cavity, Ch = chorion, Co = cotyledons, E = embryo, UC = umbilical cord, UW = uterine wall, YS = yolk sac.

her releasing prostaglandins, destroying the corpus luteum and aborting the pregnancy. The blastocyst probably achieves this by secreting oestrogen within 2 weeks of fertilisation.

Implantation, the close attachment of the foetal membranes to the uterine wall, begins about 25 days after fertilisation and is complete by about 40 days. The outer layer of foetal membranes, the *chorion*, becomes attached to the cotelydons of the uterine wall. These specialised points of attachment develop a very rich blood supply and there is some erosion of maternal tissue. The cotelydons are designed therefore to facilitate exchange of nutrients and waste products between the maternal and foetal circulations *but there is no direct mixing of foetal and maternal blood*. Foetal blood vessels develop in the chorionic wall to carry blood between the foetus and the cotyledons via the umbilical arteries and vein (Fig. 3.4 (c)). Implantation is a vulnerable stage of pregnancy. If it fails, the cow is likely to return to oestrus having missed at least one cycle (and probably more).

Twinning usually occurs because two ova are shed and fertilised. These are *dizygotic* twins and no more or less alike genetically than any two offspring from the same mother and father. *Monozygotic* or identical twins occur very rarely, perhaps once in a thousand births, when a single ovum develops two embryos. At the point where the two chorionic membranes come into contact, the two sets of foetal blood vessels unite so creating a common circulation for the two calves. If both calves are of the same sex, as is the case for 50% of dizygous twins and 100% of monozygous twins, this is no problem. However, when calves which are, genetically, male and female share the same circulation the sexual development of the female is distorted. A female calf born twin to a male is called a *Freemartin*. The developing testis of the male calf secretes an inhibitory substance that prevents its own primordial female organ, the Mullerian duct, from developing into a female reproductive tract. Unfortunately this Mullerian duct inhibitory substance enters the circulation of the female calf and inhibits its development to a degree that depends on the state of development achieved before fusion of the two circulations. Very often the vulva and external reproductive organs of the Freemartin appear normal (although the clitoris may be enlarged) but parts or all of the cervix, uterus and ovaries are missing.

Placental transfer and foetal nutrition

The close arrangement of the blood vessels of the maternal and foetal circulatory systems in the cotyledons of the placenta ensures the exchange of nutrients, water, electrolytes and the respiratory gases between mother and offspring. Water, most electrolytes, oxygen and carbon dioxide in solution pass freely between the two blood streams by simple diffusion. The principal nutrients for the foetus, glucose and amino acids, are actively transferred by specific transport mechanisms. Water-soluble vitamins and some minerals such as iron and copper are also actively transported to the foetus. There is almost no transfer of whole proteins or the fat-soluble vitamins A, D and E. The calf is therefore born lacking immunoglobulins carrying antibodies to infectious disease and deficient in fat-soluble vitamins. Both of these are normally acquired by drinking colostrum. In the case of immunoglobulins, this must occur shortly after birth while the epithelium of the digestive tract is still permeable to entire proteins.

Gestation length

The average gestation period for a Friesian cow is approximately nine months or, more precisely, 281 days, when both sire and dam are Friesians. However, sire breed affects gestation length and this has important implications for the welfare of mother and calf. Table 3.1

Table 3.1 Effects of bull breed and size on gestation length and calving problems in Friesian cows

	Gestation length (days)	Calf size (score*)	% Dystocia		Calf mortality (%)
			Cows	Heifers	
Friesian	281	3	2.7	5.7	2.4
Aberdeen Angus	279	1	–	1.4	5.3†
Hereford	282	2	1.2	2.7	2.3
Charolais	284	5	3.4	6.7	4.7
Limousin	287	3	2.4	3.2	3.3

* Calf weight is ranked on a 5-point scale from 1 = very small to 5 = very large
† Mortality in Aberdeen Angus × Friesian calves refers to calves born to heifers only

presents average values for gestation length in Friesian cows mated to bulls of different breeds, together with the incidence of calving problems (dystocia) and calf mortality.

The most popular French beef bulls, the Charolais and Limousin, prolong gestation by about 3 and 6 days respectively. However, prolonged gestation does not necessarily imply a much bigger calf. Charolais × Friesian calves tend to be very big at birth and this undoubtedly increases the risk of calving problems, especially for male calves – the incidence of dystocia for male and female Charolais × Friesian calves being 7.9% and 2.2% respectively. Limousins, on average, tend to cause fewer calving problems despite the prolonged gestation, mainly because Limousin bulls (and therefore their calves) are smaller than Charolais. It must be added that bulls within breeds differ in the calving problems they invoke in a predictable way and this can be taken into account in devising a breeding strategy (Chapter 11).

Parturition

The exact trigger for parturition (calving) is still not known for certain. It appears that the calf itself decides when it is ready (which is why bull breed influences gestation length). A trigger in the brain of the calf stimulates the release of *adrenocorticotropic hormone* (ACTH) from the anterior pituitary which in turn triggers the release of steroid hormones from the calf's own adrenal cortex. These corticosteroids cross into the maternal circulation and stimulate the placenta to produce $PGF_2\alpha$. This inhibits progesterone secretion, directly stimulates uterine con-tractions and sets in train the sequence of events that lead up to calving. This fundamental knowledge has been eminently applicable; corticosteroids and $PGF_2\alpha$ are both used in veterinary practice to initiate parturition artificially.

Blowey (1985) has described the process of calving in great detail with clear, practical advice on when and how the stockman can help and what conditions call for the specialist skills of the veterinary surgeon. Here I shall describe only the sequence of events in a normal calving.

In the last ten days before calving the cow's udder becomes very distended and milk may indeed run from the teats, particularly if she continues to be with the other cows who are regularly being milked. The secretion in the udder changes in colour and consistency from a sticky, clear honey-like fluid to the normal rich colour and consistency of the colostrum or first milk. About 48 hours before calving the ligaments of the pelvic canal relax under the influence of the hormone *relaxin* to increase the circumference of the birth canal. This can be recognised by the stockman as a loss of tension and 'dropping in' of the ligaments that run from the spinal column at the tail-head to the pin bones (Tuber ischii).

The first stage of parturition begins with contractions of the smooth muscle of the uterus induced by the hormone *oxytocin* from the posterior pituitary. The front feet of the calf are pushed towards the cervix, stretching it and sending back signals to the brain which responds by increasing the force and frequency of contractions. The cervix begins to dilate, releasing a thick, slimy substance which was the plug that blocked it during pregnancy. The cow shows some signs of discomfort at this time; she is restless and may turn to look at her flanks but she does not yet strain to expel the calf using her abdominal muscles. The exact duration of the first stage of parturition is difficult to assess because its onset is so gradual. In cows 4–6 hours would be normal and 8–9 hours not excessive. Heifers may appear to be restless and in discomfort for 2–3 days before calving.

The second stage of parturition begins when the pressure imposed by the calf on the birth canal (cervix and vagina) causes sufficient pain to induce labour – strong, controlled contractions of the abdominal muscles. The first foetal (chorioallantoic) membrane usually appears as a bladder-like protrusion from the vulva, then ruptures releasing copious quantities of allantoic fluid – the first 'bag of waters'. The calf is propelled feet first into the vagina, still at this stage covered in the amniotic membrane. The front feet, covered in protective 'golden slippers', appear at the vulva while the head and shoulders (the widest section of the calf) are still in, or just behind the pelvic canal (the narrowest section of the cow). It is normal for there to be some delay at this stage while full relaxation of the birth canal takes place. The cow may stand up or lie down according to how she feels.

When the head enters the vagina the cow usually lies down and completes the expulsion process very rapidly with a few massive abdominal contractions. The second (amniotic) foetal membrane may break during birth. If not the calf will normally struggle and rupture it successfully within a few seconds. If it doesn't, it requires assistance. The umbilical cord ruptures quickly and easily in response to movements by calf or mother and the umbilical vessels recoil and contract elastically to minimise loss of blood. It should perhaps be restated that all blood in the umbilical vessels belongs to the calf.

Normally the second stage of parturition lasts 1–2 hours. As a general rule, if a cow has been in second stage labour for 3 hours without producing a calf, she should receive veterinary or equally expert attention. This does not necessarily mean assistance. She may not be ready, but only an expert can tell.

The third stage of labour involves the expulsion of the afterbirth (placenta and foetal membranes). In normal circumstances this occurs without difficulty within 1–6 hours of calving. If left to herself the cow will probably eat the afterbirth. This was an obvious safety measure in the wild since it made the presence of a new-born calf less obvious to predators. I know of no good reason for supposing that eating the afterbirth has any important psychological or nutritional significance for the cow and very occasionally it may cause her to choke. On balance, it is better to take it away.

Lactation

Anatomy of the udder

The cow's udder consists of four entirely separate mammary glands or 'quarters', each with its own system of secretory cells and ducts and with its own storage cistern. Unless the udder is very severely damaged there is no connection between the quarters. The cells that secrete milk form a single layer around the alveoli which are microscopic sacs that empty into small and then major ducts (Fig. 3.5). These ducts lead to a series of cavities or sinuses which form a cistern or reservoir for milk. Secretion of milk by the epithelial cells of the alveoli is a continuous process. Milk is retained within the gland cisterns by sphincters at top and bottom of the teat canal. Normally, during suckling or artificial milking only the milk contained within the sinuses is actually expelled. However, the natural stimulus to milk ejection produced by bunting and sucking of the calf or the conditioned stimulus to ejection invoked by the gentle and pleasant routine in a good milking parlour induce the cow to secrete oxytocin. This causes specialised *myoepithelial cells* around the alveoli to contract and so expel milk into the duct system. This 'let down' mechanism is essential to ensure proper milk ejection which is why it is so important to preserve a gentle, undisturbed rhythm in the milking parlour.

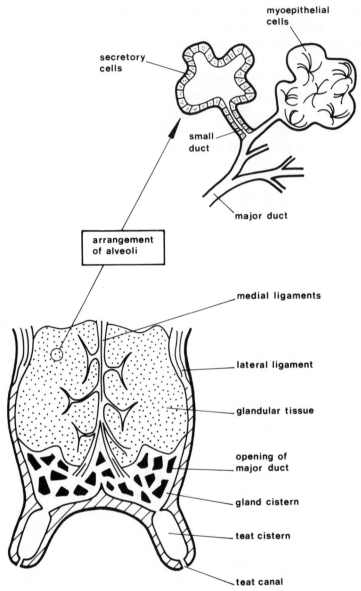

Fig. 3.5 Anatomy of the bovine udder.

Synthesis and secretion of milk constituents

The main constituent of milk is, of course, water which ranges from 850 g/kg in a typical Jersey to 870 g/kg in a typical Friesian. Other major constituents were listed in Tables 1.1 and 1.2. The osmotic concentration of milk is, of necessity, the same as that of plasma. The osmotic concentration of milk in the healthy udder is largely determined by the presence of lactose, the milk carbohydrate. In practice, this means that the ratio of lactose to water is fixed within very narrow limits; it is fruitless to contemplate any experiments designed to alter this ratio, however attractive it may appear to the producer or consumer.

Lactose is synthesised in the alveolar cells exclusively from glucose. The main milk proteins are called caseins and are synthesised only in the mammary gland from single amino acids carried to the gland in arterial blood. Casein coagulates in the abomasum of the calf or in the presence of acids and enzymes such as pepsin or rennin which separate milk into curds and whey. Casein therefore enters the clot or curd. The whey proteins in normal milk, lactalbumin and lactoglobulin, are also synthesised for the most part in the alveolar epithelium. However, immunoglobulins, the immune proteins containing antibodies to infectious agents and other antigens experienced by the cow, are synthesised by specialised plasma cells in the mammary gland and elsewhere and transported whole in the blood plasma to and through the alveolar epithelium and into the milk and especially the colostrum. The synthesis of milk protein is metabolically linked rather closely to lactose synthesis within any one genotype and is thus indirectly but closely related to water secretion. This means that while it is possible, in theory, to manipulate the protein concentration of fat-free milk by nutritional means, it is unlikely to be practicable. In other words, if one wishes to feed to produce more milk protein one must be prepared to accept more lactose and water as inevitable accompanists. If for dietary or other commercial reasons one wishes to manipulate these things, it makes most sense to do so after the milk has left the cow.

The lipid content of cow's milk is mostly true fat or triglyceride. The two major fatty acids making up these triglycerides are palmitic acid ($C_{15}H_{31}COOH$) which is fully saturated with hydrogen, and oleic acid ($C_{17}H_{33}COOH$) which is mono-unsaturated. The concentration of polyunsaturated fatty acids such as linoleic acid ($C_{17}H_{31}COOH$) is low (unless rather bizarre diets are fed). Milk fats can be synthesised from VFA (especially acetate) produced by fermentation in the rumen or longer-chain fatty acids arising either from the diet or from the fat reserves in the body. Fat synthesis in the mammary gland and its secretion into milk takes place quite independently of the synthesis and secretion of lactose, protein and water so it is relatively simple, by nutritional means, to manipulate fat concentration.

The *quantity* of milk secreted is, of course, primarily determined by the size of the mammary gland and the number of secretory cells in the alveolar epithelium, and these are under genetic control. Within this absolute limit, the quantity of milk secreted is determined by the

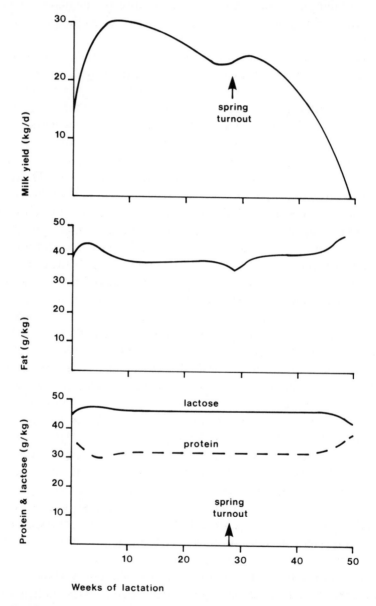

Fig. 3.6 Changes during lactation in milk yield and composition in an autumn-calving Friesian cow.

amount of nutrients carried in the blood supply to the mammary gland. This is itself dependent on three things:

(1) The supply of nutrients from the diet.
(2) The supply of nutrients arising from body reserves.
(3) Hormonal control of (a) mobilisation of body reserves, (b) partition of dietary nutrients between the mammary gland and other body tissues (such as fat).

The normal course of events for an autumn-calving dairy cow in the UK is illustrated in Fig. 3.6.

Milk yield rises to a peak by about the fifth week of lactation, persists for 3–4 weeks and then begins a slow decline. This decline persists through the winter but is interrupted by a rise when cows are turned out onto spring grass, for reasons that are both nutritional and hormonal. Assuming that the cow has been successfully re-inseminated at about the 90th day, the hormonal changes associated with pregnancy will then accelerate the decline in milk yield to the point where she can be dried off after about 305 days. Wood (1976) has developed a series of equations to predict the shape of the lactation curve for cows calving at different months of the year and much use is made of these by organisations such as the UK Milk Marketing Board to estimate future yields of individual cows or the whole herd. More simply, one can, as a rule of thumb, say that peak yield is about 0.53% of total lactation yield for autumn calvers and 0.59% of total lactation yield for spring calvers. Thus:

Peak yield (kg/d)	Estimated total lactation yield (kg)	
	Autumn calvers	Spring calvers
20	3775	3390
30	5660	5085
40	7550	6780

Moreover, one can also anticipate that the rate of decline in milk yield after about 90 days of lactation is 1.5–2.0% per week. If the decline is too rapid and total lactation yield does not come up to expectation from peak yield then something is amiss (probably nutrition). If the rate of decline after 90 days is less than 1% this suggests either that the cow isn't pregnant or peak yield wasn't what it might have been.

Colostrum contains about 780 g/kg water; it is high in protein (150 g/kg) because of the immunoglobulins, reasonably normal in terms of lipid (40 g/kg), and low in lactose (30 g/kg). This rapidly changes to the more typical composition of milk which remains relatively constant throughout most of lactation (see Fig. 3.5). However, there is typically a decline in fat concentration when cows are turned out onto lush spring grass and invariably an increase in fat concentration (and, to a lesser extent, protein) in the last few weeks of lactation when yields are low.

Hormonal control of lactation

The hormonal mechanisms for the control of the development of mammary tissue during late pregnancy and the synthesis and secretion of milk after the calf is born are extremely complex. It is necessary only to make three rather obvious but extremely important points:

(1) To ensure proper mammary development and lactation, *all* the hormone systems of the body need to be in good working order; there is no single or simple controller.
(2) The optimal hormonal balance for mammary development and lactation are not the same, thus the hormonal consequences of advancing pregnancy conflict with the hormonal requirements for peak lactation.
(3) There is competition between the mammary gland and other body tissues like muscle and fat for nutrients absorbed from the gut and this is under hormonal control.

The two hormones most responsible for controlling the partition of nutrients between the mammary gland and other body tissues are *insulin* and *growth hormone*, or *somatotrophin*. Very simply, insulin stimulates the net uptake of nutrients into body tissues like muscle and fat, and growth hormone stimulates their net release from body reserves into the circulation. In early lactation the high-yielding dairy cow secretes large amounts of growth hormone and relatively small amounts of insulin which (a) mobilises body reserves so that she can 'milk off her back', and (b) preferentially directs absorbed nutrients towards the milk-synthesising cells in the alveoli. One of the most fundamental differences between the lean, high-yielding Holstein and the fat, low-yielding Hereford is in the balance between insulin and growth hormone. Inject insulin into a Holstein in early lactation and it will produce less milk but lay down body fat – become, in fact, more like a Hereford.

While the use of hormone therapy to turn a Holstein into a Hereford (or vice versa) is a rather futile fantasy, it is important to know that insulin secretion (and the consequences thereof) can be altered by manipulating the diet or the processes of rumen fermentation. For example the antibiotic monensin sodium is an excellent growth promoter for beef animals, partly because it increases the ratio of propionate to acetate production by the rumen microbes. Increasing propionate absorption increases insulin secretion and so increases the uptake of nutrients into muscle. However, for a dairy cow this is the exact opposite of what is needed; monensin reduces both milk yield and butterfat concentration.

Synthetic bovine growth hormone or somatotropin (BST), genetically engineered in bacteria and administered to cattle by regular subcutaneous injections can increase milk yield by 5–30% depending on circumstances. For most well-fed cattle increases of 15–20% are to be expected. BST does not give farmers something for nothing. It directs a greater proportion of nutrients towards milk synthesis. This means the cow must eat more or lose condition. In early lactation the genetic capacity of the high-yielding dairy cow usually exceeds her capacity to eat, digest and absorb nutrients in her food. Administration of BST early in lactation is therefore likely to accelerate loss of body condition and predispose to infertility (Chapter 12). Administration after 100 days, when the cow is back in calf and nutrient supply has caught up with nutrient demand, can produce a sustainable increase in milk yield which then follows the normal lactation curve (Fig. 3.6) at an output (say) 15% above normal. It is therefore undeniable that BST *can* be used to boost milk production. Whether it *should* be used is a topic I shall discuss in the final chapter.

Reproduction in the male

The role of the bull in the life of the average dairy cow is, at best, a brief and infrequent affair and in most cases his direct involvement has been usurped by the middle man in the form of the artificial inseminator. Our consideration of reproduction in the male can therefore be (almost) as brief. The reproductive tract of the bull is illustrated in Fig. 3.7. The testes are formed in the abdomen during foetal development

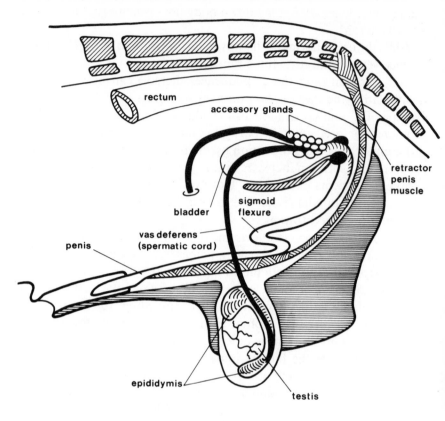

Fig. 3.7 The reproductive organs of the bull.

and descend through the inguinal canal into the scrotum before birth. They consist of seminiferous epithelium and tubules wherein the spermatozoa are made, and interstitial cells responsible for secretion of the male sex hormone *testosterone*. The hormones from the anterior pituitary that control sexual function in the male are the same as in the female. Follicle stimulating hormone (FSH) stimulates spermatogenesis and luteinising hormone (LH) regulates testosterone secretion. Puberty, as determined both by libido and the presence of mature sperm in the ejaculate, occurs at about 10 months of age, although as with the female this depends very much on growth rate.

Spermatozoa leave the seminiferous tubules at an immature stage and enter the epididymis where they are nourished by epididymal secretions and allowed to mature normally for a period of about 10 days. The most obvious feature of maturation is that the sperm acquire motility, the ability to swim by moving their tail. There is a continuous movement of sperm through the epididymis and into the vas deferens or spermatic cord. The quantity of sperm maturing in store outside the testes of the bull at any one time is about $5–6 \times 10^{10}$, which is about 10 times daily sperm production. This maturation process is essential for proper fertility; if a bull ejaculates too often not only will numbers of sperm per ejaculate fall, so too will their capacity to fertilise.

At the time of ejaculation spermatozoa are propelled from each epididymis and vas deferens into the urethra where ejaculate is augmented by secretions from a series of accessory glands which act as (a) a transport medium (b) a source of nutrients, and (c) a medium for buffering the end products of sperm metabolism and so prolonging their active life.

The penis of the bull is a long, thin structure containing much fibroelastic tissue which makes it firm even at rest. Normally it is retained within the prepuce by the retractor penis muscle which folds it into a sigmoid (s-shaped) flexure (Fig. 3.7). During sexual excitement sinuses within the penis become engorged with blood but erection is achieved mainly and very rapidly by relaxation of the retractor penis muscle which allows the sigmoid flexure to straighten.

The long, thin structure and simple erection mechanism of the bull's penis make intromission simple and rapid. Ejaculation usually occurs within 2–3 seconds of intromission with the tip of the penis close to the vaginal end of the cervix. The quantity of semen ejaculated is only about 5–8 ml but with a high sperm density (about 6 million sperm per ml).

Stockmen are well aware (or were, when herd bulls were common-place) that dairy bulls such as the Friesian are more sexually aggressive, and therefore more difficult to handle, than beef bulls like the Hereford or Aberdeen Angus. At first sight this may simply appear to be a consequence of environment; a Friesian bull tends to be kept isolated in an individual pen whereas a Hereford bull may run for much of the year with bulling heifers or pregnant cows. In these circumstances one might reasonably expect the Friesian bull to be less favourably disposed to the world at large. However, there is more to it than this; dairy bulls like the Friesian are genetically more sexually potent than the Hereford. They produce more sperm, they are more quickly aroused and they will copulate more often before exhaustion. A Friesian bull might reasonably be expected to make about 20 mounts in a period of six hours before showing signs of sexual exhaustion and 80 mounts in 24 hours is not uncommon. However, as indicated earlier, after more than 10–12 natural services per day fertility starts to fall.

4 Environmental Physiology and Behaviour

The optimal environment is one which ensures comfort, hygiene and satisfaction for the cow and optimal efficiency of production for the farmer. This simple definition can be expanded for the purpose of proper analysis as follows.

Comfort
(a) Thermal. The environment must not be so hot or so cold as significantly to effect production or cause distress.
(b) Physical. The space available to the cow and the ground or floor, walls, fixtures and fittings with which she makes contact should be such that she can rest in comfort and without interference and go about her business without risk of injury.

Hygiene
A hygienic environment is one which minimises the challenge to the animal from infectious organisms and other potentially harmful substances in the air and on surfaces such as bedding, floors, milking apparatus, etc. The environment should also not impair the defences of the animal to infections or other causes of ill-health.

Satisfaction
Each cow should be able, without undue hindrance from man or other cows, to act to satisfy her reasonable needs for food, water, comfort and social (if not sexual) intercourse, and to avoid potential or real sources of environment stress.

Efficiency of production
The environment should not prevent the cow from achieving her genetic potential for production by reducing milk yield or quality, impairing reproduction, reducing the quality and availability of food or causing ill-health.

There are two ways in which any animal responds to a change in its environment – physiological and behavioural. To take a simple example: if a cow is too hot she will sweat. Evaporation of sweat increases the rate of heat loss from the skin. In response to a change in the external environment the cow has *unconsciously* adjusted a physiological function (sweating rate) to maintain constant one feature of the internal environment, in this case, body temperature. This is an example of the mechanism of homeothermy which is itself an example of the general principle of homeostasis, the maintenance of a proper equilibrium for all the essentials of life: temperature, nutrient supply and demand, water and electrolyte balance, etc.

Suppose, however, the cow begins to feel too hot because she is standing in the sun, and shade is readily available. Rather than stay in the sun and sweat it out, she will move into the shade and so modify the environmental stimulus. This is a *conscious* behavioural response designed to ensure homeothermy (and thus comfort) with minimum effort. If the shade were a mile away she might elect to stay in the sun and sweat because it is the more efficient option. If moving to the shade took her away from food and water, or towards predators, then she might stay in the sun because she deemed these other environmental pressures to be more important.

The physiological and behavioural responses of an animal such as a cow to the many and various environmental stimuli are usually directed towards achieving the best compromise in terms of personal comfort and satisfaction. There are exceptions of course: the cow will submit herself to considerable personal discomfort or risk to nourish and protect her calf.

Physiological responses such as sweating or shivering are, for the most part, controlled subconsciously through the automatic nervous system and endocrine systems. Behavioural responses are mostly conscious, i.e. they are motivated by the animal's perception of its environment, both external and internal. For example, the child that wants an ice-cream because it sees the ice-cream van, or the cow that eats a fallen apple *because it is there*, are both motivated by a stimulus from the external environment. The cow, or child, that is shut up out of sight of food is motivated to eat in the absence of the appropriate external stimulus by hunger, a stimulus arising from the internal environment.

Thermal comfort

Heat exchanges

Homeothermy, or the maintenance of deep body temperature at about 38.6°C or 101.6°F, is achieved by balancing heat production in metabolism (H_p) against heat loss to the environment (H_l). The heat balance equation may be written:

$$H_p + H_s = H_l = H_n + H_e \qquad (4.1)$$

Heat exchange is normally measured in watts (W, or J/sec) but units of MJ/day or kJ/kg$^{0.75}$ day are probably more useful when calculating effects of the environment on food energy requirement. For a more detailed treatment, see Monteith and Mount (1974).

H_n describes 'sensible' heat loss by convection, conduction and radiation. The subscript n is conventionally used to describe sensible heat loss because it obeys Newtonian laws, i.e. it is proportional to the temperature difference between the deep body tissues of the cow and the environment.

H_e is the amount of heat lost by evaporation of water from the skin and respiratory tract.

H_s is the amount of heat stored in the body per unit time (1 day if the units are MJ/day).

Sensible heat loss

Heat produced by metabolism in the organs and tissues of the body (the body core) is dissipated through two layers of insulation provided by the tissues of the body and the hair coat (Fig. 4.1).

Tissue insulation describes the resistance to heat loss provided by the skin and subcutaneous tissues that make up the body shell. The cow can regulate sensible heat loss through the body shell to a small extent by extending her limbs to increase surface area and to a greater extent by regulating blood flow through the superficial tissues and so regulating the amount of heat convected to the surface. Vasoconstriction and vasodilation are particularly effective at regulating heat loss from the limbs. A beef cow standing out of doors at an air temperature close to 0°C can restrict blood flow to the bony extremities of the limbs below the knees and hocks to the point where skin temperature is only

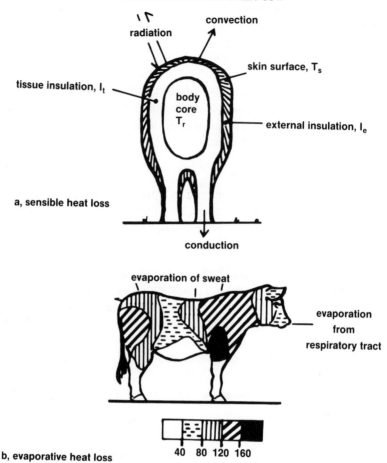

Fig. 4.1 Pathways of heat loss from cattle. (a) Sensible heat loss by convection, conduction and radiation from the body core (T_r) through two layers of insulation, tissue insulation (I_t) and external insulation (I_e). (b) Evaporative heat loss from the skin and respiratory tract. The shading indicates regional differences in maximum sweating rate from below 40% to above 160% of the average. (From Webster, 1985.)

just above air temperature and heat loss from the extremities is minimised. When air temperature falls far below 0°C, as occurs in the northern latitudes in continental America and Europe, then some cold-induced vasodilation must occur to prevent the limbs from freezing.

Maximum tissue insulation (I_t) in cold environments is described by the following equation:

$$I_t = (T_r - T_s)/H_n \qquad (4.2)$$

where T_r is rectal temperature (°C) and T_s is average skin temperature

(°C). If H_n is measured in $MJ/kg^{0.75}.d$, units of I_t are $°C.kg^{0.75}.d/MJ$. Typical values for I_t (max) are shown in Table 4.1.

Table 4.1 Heat and cold tolerance of temperate and tropical cattle acclimatised as appropriate to hot or cold conditions

	Temperate (*Bos taurus*)		Tropical (*Bos indicus*)
	Friesian	Hereford	
Heat production (thermoneutral, $kJ/kg^{0.75}$ per d)			
maintenance	500	450	400
peak lactation	900	700	620
Tissue insulation* (max)	39	46	25
External insulation† (max)	64	74	46
Evaporative heat loss (max, $kJ/kg^{0.75}$ per d)			
cutaneous	280	200	400
respiratory	80	80	80
Lower critical temperature (°C)			
maintenance	−15	−15	+10
peak lactation	(−30)	(−25)	(−5)
Limits of optimal productivity (°C)			
lower	−5	−10	+10
upper	+25	+20	+32

* insulation units are $°C.kg^{0.75}.d/MJ$
† in still air (0.2 m/sec)

The dairy cow has a special problem in cold environments because the udder is outside the body cavity and therefore particularly susceptible to heat loss especially when the cow lies on cold, uninsulated floors. Local chilling of the udder reduces local blood flow which leads to a drop in milk production because less nutrients are carried to the alveolar epithelium and *may* predispose the mammary gland to mastitis.

External insulation describes the resistance to sensible heat loss from the skin to the air provided by the coat of hair and (mainly) the layer of warm air trapped in it and on it. When the cow is standing up (Fig. 4.1), H_n mostly passes from skin to air by convection and radiation. When she lies down much heat is lost by conduction and the insulation of the surface upon which she lies assumes major importance.

External insulation (I_e) is described by an equation similar to that for I_t:

$$I_e = (T_s - T_a)A/H_n \qquad (4.3)$$

where T_a is air temperature. Typical values for I_e in still air (wind speed 0.2 m/sec) are also given in Table 4.1.

The main animal factor affecting external insulation is, of course, the depth and thickness of the hair coat. Cows of the traditional beef breeds outwintered on a hillside in Scotland or on the North American prairie clearly have thicker, denser coats than high-producing dairy cows wintering in a cubicle house and, by this criterion, greater insulation against sensible heat loss. The difference in coat thickness is partly genetic – Scottish Highland cattle are an extreme example of cattle selected for thick coats – and partly environmental in origin. Cattle have considerable ability to adjust coat depth according to how hot or cold they sense the environment to be. Hair growth is primarily controlled by day length, thus, in autumn, growth rate increases to prepare the animal for winter. Hair shedding is, however, sensitive to temperature or, more precisely, to the sensation of cold as perceived by the individual. If the cow feels cold, shedding is delayed so the coat becomes thick and rough; if she feels warm, shedding is accelerated and the coat becomes glossy and short. Thus the housed, well-fed lactating Friesian cow has a short, sleek winter coat not just because she is a Friesian but because, on the whole, she feels comfortably warm. On the other hand the Friesian cow or heifer that eats little, perhaps because it is sick or because food is not available, develops a long, thick winter coat and retains it well into the spring. It is often said that the 'poor doer' is the last to lose its winter coat and this is why.

Environmental factors affecting external insulation are wind, rain or snow, and sun. Wind, rain and snow all reduce insulation by disrupting the layer of warm air trapped within the coat and so increase sensible heat loss at a given air temperature. Sunshine obviously increases the heat load on an animal to an extent that depends on how deeply the radiant heat from the sun can penetrate the skin. A Hereford cow with a rough coat and a light skin may absorb 80% of the heat from incoming solar radiation to a depth where it can be convected round the body in the blood stream. This is useful when the sun shines in mid-winter in the temperate and cold latitudes, but a decided disadvantage in the tropics. Tropical cattle such as the white Fulani of West Africa (Fig. 4.2(a)) have a glossy white coat over a black skin which is extremely effective at repelling solar radiation; about 40% reflects off

(a)

(b)

Fig. 4.2 Cattle in the tropical mid-day sun. (a) Fulani (*Bos indicus*) in northern Nigeria. (b) Three-quarter Friesian cattle in the Egyptian Desert.

the glossy white coat, much of the rest is absorbed by the very superficial layers of black skin and transferred out again to the environment by convection and infra-red radiation before it can penetrate to a depth where it enters the blood stream and is convected to the body tissues.

Evaporative heat loss

Cattle have two very effective mechanisms for regulating heat loss by evaporation. Firstly, they can sweat copiously and for long periods. It is, of course, not the production of sweat but the evaporation of sweat that cools the animal so that if humidity is very high or the sweat gets trapped in a long, thick winter coat, the efficiency of this mechanism is impaired. The rate of sweating is under control of the sympathetic nervous system. Tropical cattle (*Bos indicus*) have a greater capacity to sweat than cattle of the temperate zones (*Bos taurus*). Within the sub-species *Bos taurus*, dairy cattle generally have a greater capacity to sweat than beef cattle and are, by this criterion, more heat tolerant (Table 4.1). Unlike man, who becomes very sodium depleted when he sweats for a long time, cattle have the ability to regulate the proportions of sodium and potassium secreted by the sweat glands. In general, ruminants consume more potassium than they need in plant foods and therefore preferentially excrete it in sweat which enables them to tolerate prolonged heat without becoming sodium depleted.

The second mechanism for regulating evaporative heat loss involves what is usually called 'thermal panting'. When cattle feel hot they increase the rate but reduce the depth of respiration so as to increase air movement over the turbinate or 'scroll' bones in the nose without increasing pulmonary ventilation. The turbinate bones, with their large surface area and copious blood supply, act as very efficient heat exchangers particularly for the blood supplying the brain. In a cool or cold environment a cow at rest will breath deeply at about 18–20 respirations per minute. The hotter the cow feels the more she will increase respiration rate up to a maximum of about 180/min. Respiration rate therefore provides a very good indication of whether a cow feels hot, warm or cold. A rate of 30–60 indicates that she is thermally comfortable. Above 60/min and she is beginning to feel hot. In general, thermal panting is used as a second resort to increasing evaporative loss. A Fulani cow at 35°C sweating copiously into a thin, glossy coat and losing heat by evaporation at $400 \, \text{kJ/kg}^{0.75}$ per day may have a respiration rate of only 35/min because evaporative heat loss from the skin is successfully preventing a rise in deep body temperature (see Table 4.1). A Hereford cow with a thick, rough coat and only managing to lose 200 kJ/kg$^{0.75}$ per day by sweating in the same environment may increase respiration rate to about 180/min and still fail to dissipate all the heat produced in metabolism.

Metabolic heat production

If a cow is neither too hot nor too cold, its heat production is determined only by the amount of work it has to do in metabolism, i.e. is independent of air temperature. In these circumstances the animal is said to be within the zone of *thermal neutrality*. It has been shown already that thermoneutral heat production is related to metabolic body size ($W^{0.75}$) which is itself closely related to body surface area. Since H_n is also related to surface area, it follows that there is little direct effect of body size on heat or cold tolerance.

There are genetic differences between cows in basal metabolism and thus in thermoneutral H_p at maintenance. The highly-tuned Friesian dairy cow has a higher H_p at maintenance than the Hereford which has itself a higher H_p than a tropical cow like the Fulani (Table 4.1). However, the major factor affecting thermoneutral H_p is food intake, because of the heat increment of feeding (see Fig. 2.10). A Friesian cow yielding 35 litres milk/day and eating accordingly may have a heat production of 900 kJ/kg$^{0.75}$/day compared with 500 kJ/kg$^{0.75}$/day when food intake is equal only to maintenance. Thus the cow that eats the most tends to be the most sensitive to heat and the least sensitive to cold.

Using figures from Table 4.1 we can illustrate the heat exchanges of a high-yielding Friesian and a dry Fulani cow at air temperatures from –20°C to +40°C (Fig. 4.3). The Friesian cow, by virtue of its high H_p, is, by this rather limited definition, at thermal neutrality at all temperatures between –20°C and +25°C. As air temperature falls so H_n increases but this is matched by a controlled reduction in H_e so that H_p is unaffected. Above 25°C, the cow is unable to dissipate 900 kJ/kg$^{0.75}$/ day of metabolic heat so must perforce reduce H_p by reducing food intake. At air temperatures close to 40°C she may be unable to dissipate even her reduced H_p, especially if the high temperature is accompanied by high humidity which reduces the evaporation of sweat. If such conditions persist the cow is no longer able to maintain homeothermy; the intensity of heat stress becomes intolerable and body temperature rises out of control.

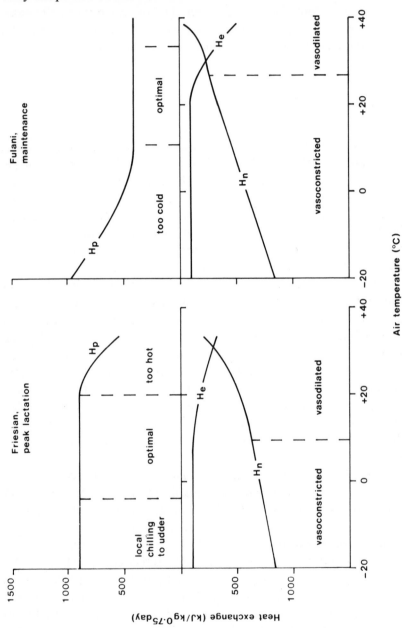

Fig. 4.3 Heat exchanges of Friesian and Fulani cattle. Hp = heat production, Hn = sensible heat loss, He = evaporative heat loss. (For further explanation see text.)

The Fulani cow, dry and at maintenance, has a much lower thermoneutral H_p (400 kJ/kg$^{0.75}$/day), and a greater capacity to evaporate heat by sweating, and can thus tolerate air temperatures up to 40°C with little adverse consequence other than perhaps a slight fall in appetite. However, her low thermoneutral H_p, I_t and I_e make her very susceptible to cold. Her lower critical temperature, which is the lower limit of the thermoneutral zone, is reached at about +10°C. Below this she has to increase H_p by shivering in order to maintain homeothermy. At –20°C her H_p has more than doubled to 900 kJ/kg$^{0.75}$/day. She would be able to sustain this for only a few days before becoming exhausted and succumbing to hypothermia.

Climate and production

Tropical climates

Heat stress in the topics varies in intensity, decreasing when the sun goes down, and it can also be ameliorated by cheap and simple measures such as the provision of a sun shade (see Fig. 4.2(b)). In hot, dry conditions such as the deserts of Israel, the Arab countries or Southern California, purebred Friesian and Holstein cows can achieve excellent yields when daytime temperatures approach 40°C provided they get shelter from the sun, high quality food, unrestricted access to water and the chance to cool off at night. In hot, humid conditions, where evaporative heat loss is depressed and the nights are almost as hot as the days, then performance is more likely to be affected. As indicated earlier, the most economically important response to heat stress is a decline in appetite to the point where the cow fails to take in enough nutrients to sustain milk yield. However, even where appetite is sustained, prolonged heat stress tends to reduce milk yield by decreasing the secretion of the lactogenic hormones. The reduction in yield involves all the major constituents but milk fat appears to be the most severely depressed.

Figure 4.2(b) shows dairy cattle feeding in the mid-day sun at a government farm reclaimed from the Egyptian desert. The cattle in the foreground are $\frac{3}{4}$ Friesian, $\frac{1}{4}$ indigenous stock. These cattle are extremely well-adapted to their environment and average lactation yield is about 5000 litres/year. The high productivity of the Friesian or Holstein cow compared with the native breeds of (say) Africa and India has provoked more or less well-meaning entrepreneurs to upgrade or substitute native breeds with Friesian/Holsteins in an attempt to increase milk production in the tropics. When this has ended in failure (as has often been the case) it has been claimed that Friesians can't perform in the heat. Figure. 4.3 illustrates why high-performance European-type dairy cattle are more sensitive to heat than native cattle, but many of the failures to farm Friesians in the tropics can be attributed to other factors: disease, poor husbandry and, especially, failure to provide food in sufficient quantity and of sufficient quality to sustain high yields. Heat intolerance is *not* the major limiting factor to the use of high-yielding dairy cows in the tropics. The exact mix of European and native breeds appropriate to a particular climate, environment and a particular local food supply is a matter of fine tuning. This book covers some of the general rules that govern specific choices but attempts no general recommendations for breeding for the tropics.

Cold climates

The lower critical temperature of a Friesian cow in still air is $-20°C$ or lower (Fig. 4.3). Factors such as wind, rain and snow, which reduce external insulation, can elevate lower critical temperature, in extreme cases to about $+8°C$. In other words, a cow in a rainstorm at $+3°C$ would shiver to elevate heat production (and so 'waste' food energy) whereas a cow in still air at $-15°C$ would not. Thus the prime requirement of winter housing (Chapter 8) is to keep cows out of the weather, rather than in the warm. However, lower critical temperature is an insufficient criterion of cold tolerance in the dairy cow since milk yield tends to decline as air temperature falls below $-5°C$ (see Table 4.1). This is because the udder is outside the body core and so subject to the consequences of local cooling, i.e. reduced blood flow and therefore milk yield. When designing winter housing for cows in cold climates it is important to ensure that the udder is protected from excessive heat loss by (a) convection – e.g. draughty conditions where cows stand before and after milking, and (b) conduction – excessive heat loss to unbedded or uninsulated floors.

Physical comfort

The dairy cow lies down to rest for about 10 hours/day. Ideally she requires a comfortable, dry, hygienic area which she can approach without disturbance, wherein she can change position (stand up, lie down or lie on the other side) without difficulty and where she can rest in comfort and security but in social contact with other cows. Chapter 8 will consider how this can best be achieved.

At this stage we can ask the cow the question, 'what kind of bed do you prefer?'. When cows are offered a selection of bedding materials in cubicles there can be no doubt from their choice that they are motivated by the desire for comfort and that their first priority is for a good mattress rather than for warmth, even at air temperatures as low as −20°C. It is hardly surprising that cows prefer not to lie on concrete; their size and shape are such that the pressure exerted at rest on skin and bone at sites such as the knee and hock (and the shear forces exerted when changing position) are far greater than those which man would experience when lying on the same surface.

On a level surface, cows tend to lie more on their left-hand side; about 65% of the time in early pregnancy, increasing as pregnancy progresses, presumably to reduce the pressure of the rumen on other contents of the abdomen, especially the gravid uterus. However, if their lying accommodation (e.g. in cubicles) is on a slope they tend to lie with their legs downhill, presumably because this makes it easier to stand up and lie down. It has been argued that cubicles should be designed on a slope to encourage cows to lie all the same way and so

Table 4.2 Behaviour of Friesian cows on the left and right sides of two houses. On Farm B cows in the left cubicles had their feet pointing down the slope when they lay on the right side and vice versa:

	Farm A		Farm B	
Slope across cubicles		level		4.2%
Bedding	Straw	Straw + 20 mm mat	Straw	Straw
Total occupancy (%)	47	82	60	58
Time, % total occupancy				
lying left side	38	38	5	66
lying right side	19	35	74	8
standing, 4 feet in	10	12	10	9
standing, 2 feet in	33	15	11	17
Time, % total lying time				
lying left side	67	52	7	89
lying right side	33	48	93	11

reduce the risk of injuries such as trampled teats. I believe this argument fails to understand the dairy cow.

Table 4.2 summarises observations of the behaviour of dairy cows in cubicles on two farms; one on the level and one with a 4.2% slope (1 in 24) across the cubicles. All cubicles were bedded with a little straw. On Farm A, cubicles on the right side of the building also had a 20 mm thick layered mat; on Farm B, the left and right sides of the cubicles on the left and right side of the building pointed downhill. On Farm B, a slope of only 4.2% made 90% of the cows lie with their feet downhill. On Farm A, cows undoubtedly preferred the cubicles with mats. With chopped straw only as bedding, lying was divided 67% left: 33% right side. When a 20 mm mat was provided the cows, which were in early lactation, showed no preference.

Conclusions drawn from preference tests of this sort need to be interpreted with care since they do not discriminate between shades of like and dislike (e.g. a preference for a particular variety of liquorice allsort or a preference for sago over tapioca pudding are both preferences but reveal nothing absolute about the taste of liquorice allsorts or tapioca). However, the fact that cows showed no preference for left or right side on the most comfortable bed suggests that when a cow does show a preference to lie on its left side it is in order to reduce abdominal discomfort due to an unsatisfactory bed. They don't lie on the preferred side all the time because this would, in time, create undue discomfort in the limbs they are lying on and, once again, the harder the bed, the greater the discomfort. The cows on Farm B, forced by the slope to lie on one side or the other nearly all the time, must therefore have been less comfortable than those on Farm A, which suggests to me that a lateral slope to cubicles is a bad thing.

There is one other aspect of cow behaviour to be gleaned from Table 4.2. Cows on Farm A in the less favoured cubicles tended to stand with their hind legs out of the cubicles and in the slurry passage. The students (Peter Creber and David Warnes) who made this study quickly spotted that they stood this way to give them a better start in the race for a more comfortable cubicle the moment it became vacant on the other side of the building.

Environment, stress and disease

A cow, like most animals leading a reasonably normal life, comes regularly into contact with environmental agents that have the potential to cause disease. In considering, very simply, how the cow copes with these challenges, it helps to classify them under three headings:

(1) *Simple infections*; diseases such as foot-and-mouth where exposure of a susceptible animal to the infectious agent produces disease which, if nature is allowed to take its course, normally proceeds to recovery.
(2) *Physiological stress*; conditions such as heat, cold, improper nutrition, exhaustion and perhaps fear which may upset the metabolism of the cow and so provoke diseases such as ketosis or hypomagnesaemia, or reduce resistance to infection.
(3) *Multifactorial diseases*; diseases which require both the presence of a specific infectious agent and some environmental stress or injury. Many of the most common diseases, such as mastitis, fall into this category.

Resistance to infection

It is necessary now to consider the very complex matter of the body's resistance to infection very simply to explain how these may be enhanced or impaired by environment and management. For further details see Tizard (1982).

All animals co-exist with a mass of microorganisms, viruses, bacteria and fungi, which have the potential to cause disease. It is important to remember that the surfaces of the animal exposed to challenge from organisms in the external environment include not only

Table 4.3   Specific and non-specific defences to infection

	Physical	Chemical	Biological
Skin	desiccation acids	acids	inflammation
Mucous membranes of body cavities	mucus urine, tears ciliary movement	gastric acids enzymes	inflammation antibodies (immunoglobulins A and G)
Blood and tissues			inflammation, fever cell mediated immunity: (1) neutrophils in blood (2) blood and tissue macrophages humoral immunity (antibodies)

the skin but also the entire epithelial surface of the gut and respiratory tract and, to a lesser extent, the epithelium of the reproductive tract and mammary gland. Microorganisms arriving at body surfaces may be destroyed or inhibited by non-specific physical and chemical means or attacked by the specific defence mechanisms of the body, e.g. antibodies and macrophages. Examples of specific and non-specific defence mechanisms are given in Table 4.3.

The first, most important, but often forgotten form of resistance to infection is innate or genetic resistance; in this case the skin, blood or mucous membranes of a particular animal species are uninhabitable for a particular organism, perhaps because they lack essential nutrients so the organism simply starves to death. Thus man does not suffer from foot-and-mouth disease, not because he develops an immunity to it but because the organism simply cannot colonise human cells. Breeding to create absolute genetic resistance to infectious disease is attractive in theory and has been used in practice to eliminate Marek's disease in poultry. Nothing yet has emerged to help control diseases of the dairy cow but with new techniques of genetic engineering it may be possible.

Many bacteria live on the surface of the skin and inside the animal but outside the body tissues in the rumen and hindgut. These organisms do not attempt to damage or penetrate epithelial surfaces and so do not provoke any of the specific body defences. Many organisms on the skin are killed rapidly by desiccation or chemicals such as organic acids in sweat and other skin secretions. Microorganisms are also killed by acid secretions in the vagina and, of course, the stomach. The vast number of microorganisms entering the abomasum of the adult cow are effectively destroyed by hydrochloric acid provided that the abomasum is not overloaded or the normal process of abomasal emptying upset by indigestion. However, the small intestine provides a favourable environment for colonisation by a number of pathogenic organisms. If indigestion does occur, particularly in the young calf with an immature rumen, it can lead to the establishment of infection in the small intestine (and so 'scours').

The mucous membranes most regularly exposed to infectious agents and other irritants are those that line the respiratory tract. The first line of defence here is mechanical. Many inhaled particles, particularly the larger ones, are trapped in the complex airways of the nose. Those that get through may deposit either in the bronchioles or alveoli. The lining of the trachea and bronchioles contains goblet cells that secrete mucus and cells containing cilia which beat rhythmically so as to move the blanket of mucus upwards out of the lungs and into the pharynx where, in the case of the cow, it is usually swallowed. This 'mucociliary escalator' removes the majority of inhaled particles that reach the lung.

If an organism avoids the non-specific physical and chemical clearance mechanisms of the animal and reaches a vulnerable site, namely an environment where it can feed and reproduce, it then comes into contact with the specific immune mechanisms of the body. Very simply, these involve (a) cell-mediated immunity, and (b) antibody production.

Cell-mediated immunity is achieved by cells that kill and eat incoming organisms. The first to arrive are the leucocytes (or neutrophils) in the blood stream which congregate at sites of inflammation and ingest what particles they can. These are reinforced in due course by macrophages manufactured in lymphatic tissue of the reticulo-endothelial system. Some of these are carried in the blood, others are restricted to the lymphatic tissue where they were formed.

Antibodies (or immunoglobulins) are produced by specialised lymphocytes called plasma cells in response to any foreign material that is sufficiently complex in structure to be recognised as an antigen. Since antibodies circulate in the blood, this is sometimes called *humoral immunity*. However, antibodies can also localise at epithelial surfaces in the gut, lung, udder and reproductive tract. The first function of the antigen-sensitive plasma cells is to recognise a complex biological molecule such as a protein as 'self' or 'non-self' and implant that information in its memory. In essence, this involves the creation of genetic machinery to synthesise specific immune proteins that will respond as antibodies to the continued presence or subsequent reappearance of the antigen.

The three most important roles of antibodies are as follows:

(1) Opsonisation: coating microorganisms and other antigens in such a way as to render them more sensitive to phagocytosis by leucocytes and macrophages. This may occur at epithelial surfaces (e.g. in the alveoli), or in the blood stream or in lymphatic tissue.
(2) Protection of epithelial surfaces: some antibodies secreted onto mucosal surfaces of (for example) the gut have the ability to coat or otherwise alter the structure of antigens in microorganisms so as to prevent them from adhering to and therefore colonising the epithelium.
(3) Neutralisation of toxins: many organisms (e.g. *Clostridium tetanus*) damage the body mainly by releasing poisons (toxins).

Antitoxins are a special form of antibody which do not attack the organism directly but inhibit its toxic effects. Having recognised the antigen, the plasma cells have then to decide whether this foreign material should be rejected or tolerated. If the decision is to reject, then subsequent presentation of antigen will provoke a much greater, more rapid production of antibody and so accelerage the removal of antigen. This is the process whereby most natural infections lead to enhanced immunity, and which is stimulated by vaccination. If the decision is to accept a foreign material (like food!) then the immune system will continue to recognise the antigen but not respond, or, in other words, show tolerance. Sometimes the system can get it wrong which is, for example, why certain men and animals become allergic to foods that are harmless to the majority.

The immune response is illustrated in Fig. 4.4.

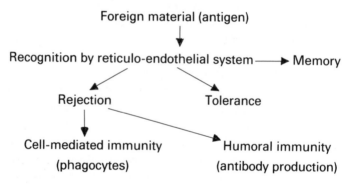

Fig. 4.4 Elements of the immune response.

'Stress'

'Stress' is a much abused word. In common (and much clinical) parlance it usually describes an environmental disturbance such as heat, cold, malnutrition, overwork, fear or emotional upset that affects the workings of the body so as directly to provoke disease (e.g. stress ulcers in man and calves) or to reduce resistance to infection. The apparent importance of stress is rooted in our folk beliefs. When a mother tells her small son who has been infected by a rhinitis virus that he has caught a chill because he, unlike his sister, has been walking around in bare feet, she is presuming that he has exposed himself to a degree of cold stress sufficient to reduce his resistance to infection significantly below that of his sister. Unsubstantiated conclusions of this kind may add guilt to the miseries of a cold, but are otherwise harmless. However, the farmer or veterinary surgeon who concludes that respiratory disease in calves is caused by chills and so blocks up the ventilators of a calf house in order to keep them warm is dangerously in error since by doing so he will massively increase the challenge to calves from infectious organisms in the air (Webster, 1985).

There is however a precise (but limited) way of describing the common physiological response of animals to stresses as various as cold, fear and overwork. This is called the *general adaptation syndrome* and has three distinct stages.

(1) The first, acute stage of the physiological response to any severe perturbation to the environment is known as the *alarm reaction*. A sudden stress such as a blizzard, prolonged calving or rough handling during loading onto a lorry will provoke an increased secretion of hormones from the adrenal cortex (*corticosteroids*) and medulla (*adrenaline* and *noradrenaline*) which mobilise the body's energy reserves in preparation for any action that may be needed to cope with that stress. At the same time the secretion of anabolic hormones like insulin will be inhibited; the body has decided that the short-term need for nutrients to do work necessary for survival overrides longer-term objectives like growth, milk production and the synthesis of immune proteins. Stresses arising from causes as distinct as cold and fear may therefore be defined as such in terms of the similar physiological response that they invoke. During the alarm phase of the response to stress there tends to be a decline in productivity and resistance to infection is impaired.

(2) If the stress is prolonged (e.g. cold stress) the animal proceeds to the stage of *adaptation*. There is still an increased secretion of adrenal hormones to sustain the continued elevation in energy metabolism but, if the stress is not too severe, the secretion of anabolic hormones returns to normal. The animal has come to terms with the environmental stress, albeit at some physiological cost. Resistance to most infections is probably normal at this time. There is also evidence that adaptation to mild stresses makes an animal better fitted to deal with a more severe physiological stress at a later stage. The incidence of ketosis in dairy cows permanently tethered in insulated barns during the Swedish winter has been shown to be seven times that in cows in loose housing with intermittent moderate exercise and cold exposure. In this case it

appears that the mild metabolic stresses of cold and exercise keep the cows in training for the much more severe metabolic stress imposed later by the onset of lactation.
(3) Stress cannot be borne for ever. Either the animal adapts to the point where the stess disappears or can at least be tolerated at an acceptable metabolic cost, or it ultimately reaches the stage of *exhaustion*. Cattle have an enormous capacity to adapt to physical stresses of heat and cold or to psychological stesses such as transport or mixing of groups. Nevertheless, chronic, severe stresses such as cold, malnutrition or severe calving problems with long-term complications can drive the mechanisms of the general adaptation syndrome to the point of exhaustion. At this stage all the defence mechanisms of the body are impaired and the cow is particularly prone to infection or metabolic disease.

As indicated earlier, many of the common diseases of dairy cows are multi-factorial in origin depending upon:
(1) The magnitude of the challenge from microorganisms and other harmful agents in the environment.
(2) The ability of these agents to reach vulnerable surfaces and tissues within the body.
(3) The competence of the specific defences of the immune system acquired either passively via the colostrum or actively by previous natural exposure to antigens, or by vaccines.
(4) Modulation of metabolism and the immune system by hormones involved in the general adaptation syndrome.

The importance of these four factors in the origin and control of specific infections and metabolic diseases will be discussed in Part III.

Behaviour

Behaviour is what an animal does. Ethology or the science of animal behaviour attempts to study what an animal does and why. A great deal of cow behaviour is reasonably obvious to an intelligent observer and one cannot, when reading publications by ethologists, avoid the impression that when they use long and strange words like 'et-epimeletic' to describe how a calf solicits the attention of its mother they are doing so to avoid giving the impression that they are stating the obvious. Nevertheless it is useful to clasify behaviour patterns of cattle if only to understand how they may be modified by the environment and how they can provide indications of normality or abnormality in terms of production, health and welfare. A good trained eye and an understanding of cattle behaviour are first essentials of the good stockman or veterinary surgeon.

The actions of any animal may be *innate* or *acquired*. When the cow licks the calf after birth and the calf struggles to its feet and seeks the teat of the cow, those are innate (i.e. instinctive or unlearned) actions. The cow that trots to the feed trough when she hears the tractor is performing an action which has been entirely acquired by learning. The month-old calf that runs across the field to its mother and immediately latches onto a teat is performing an innate action reinforced by learning. The stimuli to innate and acquired actions are external (heat, cold, predators, etc.), internal (hunger, thirst), or internal reinforced by external – much sexual behaviour falls within this category. The actions of farm animals such as cattle are directed towards:

(1) Maintenance of homeostasis, i.e. comfort and safety.
(2) Investigation of the environment, i.e. education.
(3) Establishment and preservation of security and the social order.
(4) Preservation and expansion of the family (protection of the animal's own genes), i.e. sexual and maternal behaviour.

Maintenance behaviour

This includes eating, drinking, urinating, defecating, grooming, resting and sleeping. Most of these require no comment.

EATING

This was discussed in Chapter 2. The main drive to eating is the need to replace nutrients used up in metabolism and this can properly be called *hunger*. However, the presence of palatable foods can increase food consumption by stimulating *appetite*. The main physical constraint on food intake is the quantity of undigested material remaining in the rumen. Thus a cow that fails to eat enough to meet the nutrient demands of early lactation is simultaneously hungry but full up.

REST AND SLEEP

Cows like to lie at rest for 9–12 hours in any 24–hour period. They usually lie on their briskets with both forelegs tucked under their body, the weight of the abdomen resting on one hind limb and the other hind limb stretched out to one side. As indicated earlier, they have a slight preference to lie on the left side but regularly change sides to avoid putting excess pressure on the laid-on limb. If a cow lies for several hours on one side, e.g. after an exhausting calving, this limb can become paralysed due to pressure on its nerve supply. Cows normally lie down front legs first and get up front legs last. in each case this involves a forward lunge, and cubicle design should take this into account. However, many cows in poorly designed cubicles learn alternative strategies for lying and changing position and may adopt abnormal but apparently successful postures such as 'dog-sitting'.

Because most cows in a dairy herd are busy during the day there are times at night when *all* cows wish to lie down. The number of cows in a cubicle house should not therefore exceed the number of cubicles.

Cows do experience true sleep that appears on electroencephalograph records to be identical to sleep in man. In a secure environment they sleep on average for about four hours – again, almost entirely at night, and furthermore drowse (the post-luncheon lecture syndrome) for a further six hours or so. The intensive dairy producer who elects to milk his cows three times a day and provide feed in and out of parlour five times a day is encroaching on their natural desire for rest.

Investigative behaviour

All animals have basic innate drives to learn by exploration of the environment and to seek security from learned and probably instinctive dangers. Faced by a novel object a cow or calf will smell it, lick it and, if possible, attempt to eat it. This investigative behaviour is undoubtedly instructive but can get cattle into trouble, as when they eat flaking lead paint, polythene bags, pieces of wire or even (in my own experience) old car batteries.

Social security

Cattle are a herding species, i.e. they prefer to be in groups rather than in isolation. The probable explanation of the herding instinct in non-aggressive herbivores like cattle and sheep is that it has evolved as innate activity designed to protect the individual. A cow on its own that encounters a hungry lion on the African plain has perhaps an even chance of survival. The individual cow in a herd of forty has no worse than a 1 in 40 chance of getting killed. Within the herd however, cattle seek and preserve their own social space. Cattle, even young calves, do not usually huddle as pigs do, almost certainly because they do not need to do so to keep warm. At air temperatures below 0°C calves are more inclined to huddle and at air temperatures above 25°C weaned pigs tend to spread themselves.

When two cows unknown to one another converge on the same space it is likely to provoke aggression. The situation is, of course, much worse with bulls. Cows will fight either by pushing head to head or by making swinging, gouging or lifting movements with the head into the hindquarters or flank of the opponent. When cows have horns this can be painfully effective. Most conflicts are resolved fairly rapidly when one cow elects to submit. In unstable herds where cows have not established their place in the social hierarchy it is usually the larger, heavier cow (or the cow who has horns) who wins. Thereafter each cow usually remembers its place. On subsequent encounters the cow which deems itself to be inferior assumes a submissive posture with its head held low, the dominant cow acknowledges this and unless there is severe competition, e.g. for trough space, both parties sensibly accept the status quo. In relatively small, stable herds older cows tend to assume dominance by virtue of age, newly arrived heifers adopt a submissive attitude from the outset and maintain this respect for the older animal so long as both are in the herd. It is however almost inevitable that some cows will bully some heifers or other new arrivals. It is important therefore to ensure that they sort themselves out well before critical times like calving and that heifers and other new arrivals are not prevented by other cows from getting to the food trough and to bed. A heifer that calves for the first time at grass and is then subjected to the strange environment of a cubicle house, collecting yard and milking parlour and then further subjected to aggression from older cows, can become very frightened and bewildered. She may be reluctant to approach the feed fence or cubicles and stand, or worse, lie in the dunging passage for many hours. Not only is this an insult to her welfare, it can predispose to many of the veterinary problems of early lactation, especially metabolic disorders, environmental mastitis and lameness.

It has been claimed that cows can recognise and remember about 50–70 other cows as individuals. This implies that in stable herds of 50 cows or less each cow remembers its place and aggression is minimised except when new heifers arrive and are recognised as strangers. There is some evidence that aggression on a daily basis between adult cows is greater in larger herds but provided there is good management, adequate beds and trough space, I do not think it presents a great problem. It may also be easier for a heifer to merge unnoticed into a larger herd.

Sexual and maternal behaviour

Sexual behaviour was described briefly in the previous chapter and I shall in Chapter 12 return to the important practical matter of recognition of oestrus from behavioural and physiological indicators. Dairy cows get little or no chance to demonstrate maternal behaviour except on a few farms where the calf is kept with the cow for a few days or weeks. There is an obvious case to be made for keeping the calf with its mother for the first 24 hours of life to give it the opportunity to take in as much colostrum as possible. Mother's milk is not only ideal food but also continues to confer some local immunity to the gut wall of the calf long after the epithelium has closed down to prevent the absorption of large immune proteins. In some dairy areas of the South West of England calves have traditionally been allowed to feed from their mothers for about a fortnight – and very well the calves look. The advent of quotas on milk production has encouraged other farmers to consider this as a profitable way of using up surplus milk. Nevertheless I believe that it is kinder to the cow to remove the calf as soon as possible after birth. Of course, she has innate maternal instincts but if a Friesian cow is removed from its own mother at birth and subsequently allowed no time with her own calves her maternal instincts are not reinforced by learned behaviour. The weaning of a calf from a cow (the French word *sevrage* is more accurate) at the age of two weeks is extremely distressing to cow, calf and anyone else who happens to be in earshot.

Abnormal behaviour

Abnormalities of behaviour are a subject of much concern to the farm animal ethologist, particularly in connection with those intensive husbandry systems like battery cages for laying hens and individual stalls for sows or veal calves which limit the expression of normal activities and provoke unusual behaviour patterns such as bar chewing in sows. Bar chewing may be abnormal by definition but that does not necessarily imply that it does any harm or causes any distress. Indeed, there is some recent evidence that when sows compulsively chew the bars of their stalls this releases *beta-endorphins*, natural brain opiates which make the pig feel good – or at least better.

Cattle deprived of the opportunity to perform normal oral activities do engage in a number of purposeless oral activities such as tongue-rolling (Webster, 1985). This is common in veal calves on all-liquid diets, fairly common in beef cattle on cereal-based diets, but very rare in cows given unrestricted access to roughage.

The only other abnormal oral activities of cattle are urine drinking which is potentially harmful and milk sucking which is infuriating. Urine drinking is rare in cows, although quite common in calves reared intensively for veal or baby beef. Again, I believe it is caused by not having enough useful things to do with the mouth, which is not only the organ that eats but also the one that investigates. Urine drinking is more common in males, probably as an extension of pizzle sucking. There have been several hypotheses advanced as to why individual cows choose, as adults, to suck milk from the teats of other cows but none, to my knowledge, that can be put to good use in practical husbandry. There are contraptions which can be put on the nose of such cows to prevent them from sucking but, on the whole, it is better to remove them from the herd.

Part II

Feeding the Dairy Cow

5 Nutrients and Nutrient Allowances

It is possible to feed dairy cows reasonably successfully without any understanding of nutrition, in much the same way that most mothers successfully rear their children; food is offered according to availability, palatability, cost and experience of what seems to work in practice. The farmer assesses response to food in terms of milk yield in his dairy cows, the mother in terms of the growth, health and strength of her children.

The traditional approach of the dairy farmer to feeding his cows over winter was based on the simple premise that grass conserved as hay or silage provided the maintenance ration plus a little milk, if the quality was good. Milk yields in excess of (say) 1 gallon (or 4.5 litres) were sustained by concentrate food in the form of a cereal-based cake at 4 lb cake per gallon of milk (or 0.4 kg cake per litre milk). In easier days the dairy farmer could get by with this approach, particularly if he kept a close eye on performance and adjusted concentrate feeding if the cows were doing better or worse than expected. However, that was about as far as he could go to modify or improve the system by improving yield, milk quality or food conversion efficiency through improvements in foods and feeding methods since he had no proper basis for relating input of food to requirements for lactation in terms of a common measurable currency, namely nutrients.

Foods contain nutrients, the four main categories being energy, protein, minerals and vitamins, and farm animals require nutrients for four main purposes – maintenance, growth, pregnancy and lactation.

The purpose of nutrition is to balance supply and demand for specific nutrients for specific purposes according to defined rules and conventions. The animal feeder needs an understanding of nutrition in the same way as a musician needs to be able to read music.

Practical nutrition has four main themes:

(1) Nutritive value of foodstuffs – the description and prediction of the amount of available nutrients within unit mass of food (usually expressed per kg dry matter, DM).

(2) Nutrient requirements of animals – description and prediction of the quantities of specific nutrients required for maintenance, growth, pregnancy and lactation.

(3) Ration formulation – the preparation of rations so that the available nutrients contained in food match nutrient requirement as accurately as possible, and meet nutrient requirement at least cost.

(4) Disorders of nutrition and metabolism – the diagnosis and prognosis, prevention and treatment of metabolic disorders from precise evaluation of nutrient supply and demand.

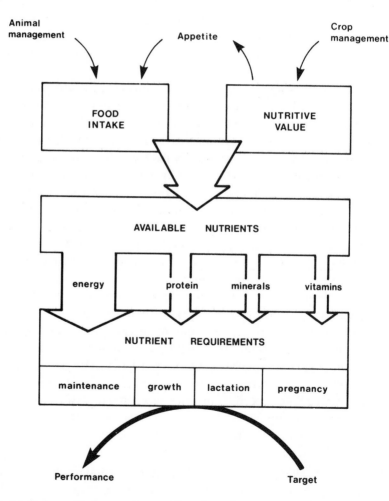

Fig. 5.1 Factors affecting nutrient supply and requirement.

Nutrient requirements and responses

The procedure for assessing the nutrient requirements of any farm animal for any productive function is illustrated in Fig. 5.1. In the case of the dairy cow, let us assume that she has, in the short term, a target of 30 kg milk/day based on her present yield and that which she achieved in her last lactation. Advisory services for dairy farmers, such as those provided in the UK by the Milk Marketing Board (MMB) or the major feed companies, predict future individual and herd yields based on past records. This, in the first analysis, constitutes the *target*. The nutritionist then draws up a list of requirements for energy, protein, minerals and vitamins to meet that target and devises a ration based usually on restricted amounts of concentrate feed and unrestricted access to grass or conserved forages so as to meet requirements for all nutrients within the constraints of appetite. Most attention is given to getting the supply and demand for energy in precise balance because energy is the nutrient required in greatest abundance and therefore the most expensive to supply. The ration is then balanced for protein, minerals and vitamins in turn.

The ability of the ration to meet the nutrient requirements of a cow for maintenance and lactation depends, of course, on the quantity provided and its nutritive value (Fig. 5.1). Food intake can be manipulated by attention to such items of management as space at food troughs or cutting or carting silage rather than relying on a self-feed system. The nutritive value of a forage crop such as silage is critically dependent on harvesting methods. Improving nutritive value can increase food intake by improving palatability or by reducing retention time in the rumen (Chapter 2).

When the supply of nutrients has been balanced as closely as possible to requirement then the response of the cow or herd can be assessed by comparing performance from milk records with targets predicted (for example) by MMB. If performance, here assessed by milk yield, fails to meet target then the farmer and his nutrition and/or veterinary advisor should proceed through the nutrients in sequence to discover where supply is failing to meet demand and whether this is due to a failure in quantity or quality of food eaten. To quote Roger Blowey, 'common things occur commonly'; poor performance is much more likely to be due to a lack of the major nutrient, energy, than to deficiency of a trace mineral, especially in dairy cows who usually receive generous amounts of the minor nutrients in compound foods. Other failures of performance, such as infertility, can be examined in the same way to discover the extent to which they may be nutritional in origin.

If performance is on or above target – or on target but above quota – then the farmer may well decide that he is happy with the status quo. Alternatively he may seek a way of reducing input costs, e.g. by substituting more forage for concentrate or home-produced cereals for compound food. When each available food is properly described in terms of cost (£/tonne) and nutritive value it is possible to calculate the combination of foods that meets nutrient requirements at least cost.

The concept of feeding according to nutritional requirement is fundamentally sound because it seeks to maximise the efficiency of conversion of animal food to animal product. I should add here that practices such as 'lead feeding', 'flat-rate feeding' and 'brinkmanship,' discussed in Chapter 6, are all variations on this theme. However, it is a concept that should not be applied too rigidly. Tables of nutrient requirement, even as modern and detailed as those of the UK Agricultural Research Council (1980) or the US National Research Council (1985) are only approximations to the average requirement of the average cow eating food of uncertain nutritive value. It is not surprising therefore that some cows and some foods perform much better or worse than expected.

There is a second, more fundamental, difference to be recognised between requirement and response. Requirements for energy, protein and minerals for (for example) lactation are calculated using the rules described in Chapter 2, from:

(1) yield of energy, protein and minerals in milk.
(2) digestibility (or availability) of energy, protein and minerals in food.
(3) efficiency of utilisation of absorbed nutrients for milk production.

This factorial approach may be criticised in that it considers nutrients in isolation. Moreover, it calculates the *least* amount of nutrient required to meet a specified requirement, e.g. a milk yield of 30 l/day. This defines the optimal biological efficiency of conversion of, say, food protein into milk protein but does not consider the possibility that increasing protein intake in excess of requirement might, for a variety of indirect reasons that I shall discuss later, increase milk yield further and therefore make more money – even though the extra protein may be used at a lower biological efficiency.

The production *response* of most animals to most nutrients is subject to the law of diminishing returns. This is illustrated schemati-

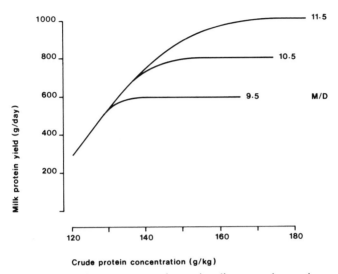

Fig. 5.2 Milk protein yield in response to increasing dietary crude protein concentration at different levels of energy intake as determined by the ME concentration in dry matter (M/D, MJ/kg).

cally in Fig. 5.2. The three curves depict increases in milk protein yield (100 g milk protein = 30 l Friesian milk) in response to increasing the protein concentration of the ration from 120 to 180 g/kg DM when the metabolizable energy content (M/D) is 9.5, 10.5 and 11.5 MJ/kg DM. If we assume as a first approximation that all animals consume the same amount of DM then M/D determines intake of metabolizable energy (ME).

When M/D = 9.5 the cow shows a production response to increasing protein concentration up to 135 g/kg. Thereafter further increases in protein produce no further increase in yield because ME supply is now limiting production and the excess protein is being used as an energy source. As energy supply is increased so the cow becomes able to respond to further increments of protein to the point where a diet that is rich in ME (M/D = 11.5) produces peak yields when crude protein concentration is 160 g/kg. Increasing protein concentration from 150 to 160 g/kg usually involves feeding protein in excess of requirement as calculated by the factorial method as described in the following section. Whether it is justified in practice depends upon the financial benefit accruing from the increased yield relative to the cost of increasing dietary protein concentration. For any nutrient and any response subject to the law of diminishing returns the cost:benefit ratio declines with each successive increment to input.

The purpose of this introduction has been to show that the proper feeding of dairy cows is based, in the first analysis, on arranging nutrient supply to meet nutrient requirement. However, this logic should not be applied too rigidly, partly because of uncertainties in

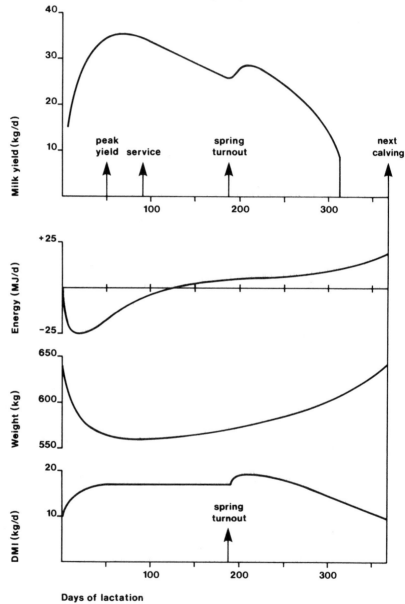

Fig. 5.3 The sequence of events during lactation for a typical autumn-calving Friesian cow.

assessing supply and requirement and partly because animals can show responses to increments of nutrients that the requirement concept cannot embrace. I shall return to these later. For the moment it is necessary to examine in detail the requirements of dairy cows for the major nutrients.

Nutrient allowances for the lactating cow

What we require of the dairy cow is that she has a good, normal lactation. Figure 5.3 illustrates simply what this involves for a typical adult Friesian cow calving in October. We expect her to proceed normally to reach a peak milk yield of (say) 35 kg/day within about 40–50 days of calving. We expect her to return to oestrus in sufficient time to ensure that she can be inseminated at about Day 90 of lactation so as to calve again at about Day 365. After about Day 100 we expect her milk yield to decline at a rate of 1.5% to 2.0% per week, this decline being interrupted (usually) by a brief rise after she is turned out onto spring grass. Thereafter as pregnancy proceeds, the rate of decline in milk yield becomes steeper and we would expect to dry her off at about 305 days to give her 2 months rest before her next calving.

Her nutrient requirements for maintenance, lactation and pregnancy have to be met within the constraints of appetite (DM intake) which may be about 10 kg/day just before calving and rise to 17 kg/day within about 40 days. During the first weeks of lactation it is normal for dairy cows to lose weight. Figure 5.3 shows first the inevitable loss of calf and other products of conception followed by a further loss of about 40 kg over the next 40 days as the cow 'milks off her back'. A Friesian cow that loses much more than 40 kg during early lactation is probably being underfed although she may be an exceptionally good milker. By the same argument, one that loses less than 20 kg is almost certainly not giving enough milk. The daily balance of energy and most other nutrients is therefore initially negative, reaches equilibrium by about 100 days and must thereafter be positive to ensure that the cow not only recovers lost tissue reserves but also provides all the nutrient requirements for her next calf.

Because the physiological capacity of the high-yielding dairy cow to direct nutrients into milk in early lactation nearly always exceeds her capacity to eat, digest and absorb nutrients, it follows that feeding the cow during at least the first half of lactation is a matter of ensuring that the nutritive value of her food, its palatability and its presentation are all such as to ensure maximum nutrient supply. From mid-pregnancy onwards it is possible for cows to eat in excess of requirements and hereafter it is important to regulate the supply or quality of nutrients to ensure that the cow does not get too fat. The calculations and tables of nutrient allowances that follow were based initially on the approach adopted by the UK Agricultural Research Council (ARC) and described in detail in *The Nutrient Requirements of Ruminant Livestock* (ARC, 1980). In the second edition these have been combined with the new proposals contained with the UK Metabolizable Protein System (Webster, 1992). Considerable reference is also made to the USA National Research Council publication *Nutrient Requirements of Dairy Cattle* (NRC, 1985).

In drawing up these tables a distinction is made between *requirement* and *allowance*. A published value for requirement means, in effect, the amount of nutrient required to meet a stated level of production in the average cow of that size and breed but does not take into account individual variation. The farmer who feeds to meet the requirement of his average cow, by definition, underfeeds approximately half the cows in the herd. An *allowance* is defined as the amount of nutrient required to ensure that nearly all his cows receive at least their requirement. In most cases allowances are calculated by increasing requirement by 5%.

Energy

In the UK energy allowances and the energy value of foodstuffs are both defined in terms of (apparent) metabolizable energy (ME,MJ). The USA system uses either ME(Mcal) or net energy for lactation (NE_l) where $NE_l = k_l\ ME$ as described in Chapter 2.

MAINTENANCE

As a first approach the ME allowance for maintenance of a dairy cow engaged in a modest amount of daily activity and not exposed to cold stress may be taken as 0.5 MJ ME/kg $W^{0.75}$ per day, thus:

	Wt (kg)	$W^{0.75}$	ME_m
Jersey	350	80.9	40
Friesian	600	121.2	61

Table 5.1 Nutrient allowances for dairy cows (based on UK recommendations)

	Breed					
	Jersey	Guernsey	Ayrshire	Shorthorn	Friesian	Holstein
Body weight	350	450	500	550	600	700
Appetite (kg DM/day)	12.5	14	15	16	17	19
Daily allowances for maintenance						
metabolizable energy (MJ)	40	49	53	57	61	68
metabolizable protein (g)	188	227	246	264	282	317
calcium (g)	12	16	18	20	21	23
phosphorus (g)	15	23	25	27	28	30
magnesium (g)	5	7	8	9	9	10
Milk composition (g/kg)						
protein	37	35	33	33	33	33
fat	49	49	38	36	37	37
total solids	140	135	125	123	124	124
Allowances for milk production (per kg)						
metabolizable energy (MJ)	6.0	5.7	5.0	5.0	5.0	5.0
metabolizable protein (g)	60	56	53	53	53	53
calcium (g)	3.2	3.2	2.8	2.8	2.8	2.8
phosphorus (g)	1.7	1.7	1.7	1.7	1.7	1.7
magnesium (g)	0.6	0.6	0.6	0.6	0.6	0.6

Other examples are given in Table 5.1. In the USA, maintenance allowances are about 12% higher (Table 5.2).

LACTATION

The ME requirement for lactation is determined by the energy value of milk (EV_l) and the efficiency of utilisation of ME for lactation (k_l). In the UK, EV_l is calculated by:

$$EV_l \text{ (MJ/kg)} = 0.0386 \text{ BF} + 0.0205 \text{ SNF} - 0.236$$

where BF is butterfat and SNF solids-not-fat (g/kg). The efficiency of utilisation of ME, k_l, is assumed to be constant at 0.62. Multiplying by 1.05 to convert a requirement into an allowance, we have:

$$ME_l \text{ (MJ/milk)} = 1.05 \ (EV_l/0.62) = 1.694 \ EV_l$$

ME_l allowances for a typical Friesian and Jersey become 5.0 and 6.0 respectively (Table 5.1).

Table 5.2 Nutrient allowances for dairy cows based on USA (National Research Council) recommendations

| | Body weight (kg) | | |
	350	600	700
Daily allowance for maintenance			
ME (Mcal[MJ])	10.6[45.2]	16.12[67.6]	18.10[76.0]
NE$_l$ (Mcal[MJ])	6.47[27.1]	9.70[40.7]	10.89[45.7]
Crude protein(g)	341	489	542
Milk fat (g/kg milk)	49	37	37
Allowances for milk production (per kg milk)			
ME$_l$ (Mcal[MJ])	1.37[5.8]	1.20[5.0]	1.20[5.0]
NE$_l$ (Mcal[MJ])	8.82[3.4]	0.71[3.0]	0.71[3.0]
Crude protein(g)	97	84	84
Recommended nutrient concentrations in DM			
Milk yield (kg/day)	17	27	35
ME (Mcal/kg[MJ/kg])	2.71[11.3]	2.71[11.3]	2.71[11.3]
Crude protein (g/kg)	150	150	150

In the USA, k_l is assumed to be 0.6 so NE$_l$ = 0.6ME but there is no correction factor; their estimate of ME$_l$ for a Friesian cow is thus exactly the same as that in the UK, 5.0 MJ/kg (Table 5.2).

Table 5.3 gives ME allowances for maintenance and lactation in Jersey, Ayrshire, Friesian and Holstein cows and the ME concentration within the dry matter (M/D, MJ/kg) necessary to meet ME allowance within the constraint of appetite. The calculations proceed as follows:

	Jersey	Friesian
Milk yield (kg/day)	17	27
ME for maintenance (MJ/day)	40	61
ME for lactation (MJ/day)	102	135
Total ME requirement (MJ/day)	142	196
DM intake (kg/day)	12.5	17.0
ME concentration in DM (MJ/kg)	11.3	11.5

Table 5.3 Typical diets and dietary concentrations for dairy cows in mid-lactation (based on UK recommendations)

	Jersey	Ayrshire	Friesian	Holstein
Appetite (kg DM/day)	12.5	15	17	19
Milk yield (kg/day)	17	22	27	35
Requirements per day				
ME (MJ)	142	163	196	243
MP (g)	1208	1412	1713	2172
calcium (g)	66	80	97	121
phosphorus (g)	44	62	74	90
Recommended nutrient concentrations in DM				
ME (M/D, MJ/kg)	11.3	10.8	11.5	12.8
calcium (g/kg)	5.3	5.3	5.7	6.4
phosphorus (g/kg)	3.5	4.1	4.3	4.7

One important point that emerges at this stage is that a typical Jersey requires food of the same quality as a typical Friesian.

ENERGY BALANCE IN TISSUES

In early lactation the cow is expected to contribute energy from her own tissues; from mid-pregnancy onwards it is necessary to feed for body gains.

It is assumed that, during lactation, the energy value of body gain (EV_g) is 20 MJ/kg and that ME is utilised for body gain with the same efficiency as for lactation, i.e. $k_f = k_l = 0.62$. Thus:

ME allowance for body gain = 1.05(20/0.62) = 34 MJ/kg

or

1 MJ ME in excess of requirement for maintenance + lactation = 0.03 kg liveweight gain

In early lactation, it is assumed that the energy in body tissues is converted to milk energy at an efficiency of 0.82. Thus:

1 kg tissues = 0.82 × 20 = 16.4 MJ milk energy

This is equivalent to

$$1.05/0.62 \times 16.4 = 28 \text{ MJ ME}$$

If ME intake is 28 MJ/day below ME requirement the cow would then be expected to lose 1 kg/day.

To illustrate this rather difficult concept, consider a cow weighing 680 kg and giving 38 kg milk/day in early lactation and eating 19 kg DM/day of cake and silage.

Food	DM intake (kg)	M/D	ME(MJ/day)
Cake	10	13.5	135
Silage	9	10.0	90
		ME intake	225
ME requirement (63 + 190)			253
ME equivalent of liveweight loss (MJ/kg)			28
Predicted liveweight loss (kg/day)			1.0

This example also provides a very simple introduction to the principle of ration formulation based, in this case, on a fixed ration of cake (10 kg DM/day) and an estimated *ad libitum* silage intake of 9 kg DM/day. All these values are approximations carrying a fair element of uncertainty. If the cow lost less than 1 kg/day it could be because:

(1) she was eating more silage than predicted;
(2) the silage, had a higher nutritive value (M/D) than predicted;
(3) the energy value (EV_g) of her liveweight loss was higher than predicted due to an increased contribution of body fat (39 MJ/kg) to weight loss.

If weight loss was considerably in excess of 1 kg/day because any or all of these factors had been incorrectly estimated in the opposite direction, then this would be a cause for concern. If this excess weight loss could not be overcome by improved feeding then the cow would be unlikely to sustain her milk yield and would be more likely to exhibit infertility.

PREGNANCY

Since growth rates of the calf and other products of conception increase exponentially throughout pregnancy, ME requirement for pregnancy is usually calculated as an exponential function of number of days pregnant (t):

$$ME_p \text{ (MJ/day)} = 1.13e^{0.0106t}$$

The demands of pregnancy are, in fact, quite small, thus:

Days pregnant	ME_p (MJ)
90	3
180	8
270	20

This should, of course be added to ME requirement for maintenance, lactation and body weight gain.

Protein

Chapter 2 described in some detail how the protein requirements of cows need to be considered in two stages.

(1) Requirement for rumen-degradable protein to meet the needs of the rumen microorganisms for protein synthesis.
(2) Requirements of the cow for amino acids, or metabolizable protein to meet the needs of the tissues for maintenance, growth, pregnancy and lactation.

The maintenance needs of the tissues for metabolizable protein are taken as 2.2 g MP/kg $W^{0.75}$ per day plus a further 0.13 g MP/Kg $W^{0.75}$ per day to replace protein losses in hair and scurf, making 2.33 in all. This factor generates the MP maintenance requirements in Table 5.1. The MP equivalents of liveweight gain and loss are 254 g/kg gain and 110 g/kg loss. It is less easy to produce a convincing figure for k_{p_l}, the efficiency of conversion of MP into milk protein. The evidence from feed trials is rather confusing, mainly because the efficiency of protein utilisation is only maximal when protein supply is inadequate. Responses to increments of protein fed at commercially sensible concentrations are subject to the law of diminishing returns as illustrated in Figure 5.2. Figure 5.4 summarises, very simplistically, my attempt to interpret practical feeding trials with dairy cows to examine

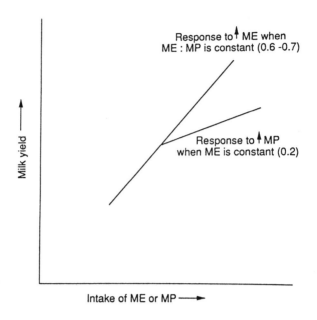

Fig. 5.4 Efficiency of response to increasing ME and MP in dairy cows.

responses to increasing ME and MP (Webster, 1992). When cows are fed diets where the ratio MP:ME is constant (at approximately 8 g MP:ME) the net efficiency of conversion of increments of MP to milk protein is between 0.6 to 0.7, essentially the same as k_l, the net efficiency of conversion of ME to milk energy. This is more than coincidental and suggests that the primary determinant of milk yield in this case is ME. When calculating MP requirement for milk production I have therefore assumed $k_{p_l} = k_l = 0.62$ (Table 5.1). However, when ME is held constant, e.g. by the constraints of dry matter intake, then the net efficiency of response to increasing MP, by increasing the ratio MP:ME is 0.2. The implications of this for the fine tuning of ME and MP to meet milk quotas are rather subtle and I shall return to them later.

The requirement of the rumen microorganisms for RDP cannot be related directly to ME supply since they can only utilise that proportion of ME which is fermentable. The requirements for microbial crude protein synthesis are defined by *Effective Rumen*

Degradable Protein (ERDP) and *Fermentable Metabolizable Energy* (FME) where

ERDP = 0.8 QDP + 1.0 SDP (see Chapter 2)

$FME = ME - (ME_{fat} + ME_{fermentation\ products})$

The ratio FME:ME is reduced when diets contain significant amounts of fats (e.g. some high energy dairy cakes) or significant amounts of prefermented material. In silages 10% (or more) of ME may be in the form of prefermented lactic or volatile fatty acids which are available as an energy source to the cow but not to the rumen microbes. Tables 5.5 and 5.6 list the nutritive value of some typical feeds for cattle and include best estimates of FME and ERDP. These values are drawn largely from the Interdepartmental Working Party Report (1993) on Metabolizable Protein.

The yield of microbial crude protein from the rumen is assumed to range from 9 g/MJ FME at maintenance to 11 g/MJ FME at three times maintenance, the increase reflecting increased outflow rate from the rumen (Chapter 2). Thus:

ERDP requirement = 9 to 11 g/MJ FME.

Consider our Friesian cow with a milk yield of 27 kg/day (Tables 5.1 and 5.3):

ME requirement = 196 MJ/day
MP requirement = 1713 g/day
Assume a diet for which FME:ME = 0.85
FME supply = 167 MJ/day
ERDP requirement (11 g/MJ FME) = 1837 g/day
MP from microbial protein: (0.64 × ERDP) = 1176 g/day
MP requirement from digestible undegradable protein (DUP) = 537

Minerals

There are several ways of thinking about minerals. The nutritionist seeks to discover what specific elements are involved in body form and function and to measure the exact quantity of each that needs to be provided in the diet. The farmer and vererinary surgeon only become concerned if requirement exceeds that which is likely to be contained in 'normal' feeds. I shall, for the most part, adopt this approach. Finally, the manufacturer of compound feeds can afford to err on the side of generosity unless the mineral is particularly expensive and provided that excessive intake does not lead to toxicity (e.g. copper) or other metabolic disorders (Chapter 7). Recommended daily allowances for the major minerals for maintenance and lactation given in Table 5.1 are based on ARC (1980). Table 5.4 presents NRC (1978) recommendations for concentrations of major minerals and trace elements in the rations of dry cows, lactating cows and growing heifers, plus typical concentrations of these elements in alfalfa and barley, two feeds in very common use. This illustrates, for example, that while cattle have substantial requirements for potassium and iron it is most unlikely that they will, in practice, ever be short of either of them. For copper and cobalt, on the other hand, supply from these typical foods only just meets average requirement which means, inevitably, that some foods will be deficient for some animals.

Table 5.4   Recommended mineral concentrations in rations and food for dairy cattle (from NRC, 1978)

	Concentrations in rations			Typical concentrations in food	
	Dry cows	Lactating cows	Growing heifers	Alfalfa grass	Barley
Major minerals (g/kg DM)					
calcium	8.7	5.0–6.0	4.0	17.2	0.5
phosphorus	2.6	3.0–4.0	2.6	3.1	3.6
magnesium	1.6	2.0	1.6	2.7	1.4
potassium	8.0	8.0	8.0	20.3	4.5
sodium	1.0	1.8	1.0	1.5	0.3
Trace elements (mg/kg DM)					
iron	50	50	50	300	90
copper	10	10	10	10	9
cobalt	0.10	0.10	0.10	0.1	0.1
manganese	40	40	40	50	20
zinc	40	40	40	20	20
iodine	0.25	0.50	0.25	–	–
selenium	0.10	0.10	0.10	–	–

Calcium is required in greatest quantity to mineralise bone so at least 98% of it is contained within the skeleton. The other 2% is concerned primarily with the initiation and maintenance of nerve and muscle function. Thus acute calcium deficiency is associated with neuromuscular failure ('milk fever', Chapter 7); chronic deficiency is associated with bone weakness or osteoporosis. Grasses are good sources of calcium; alfalfa (or lucerne) and brassicae (e.g. kale and rape) are particularly rich; cereals and roots tend to be poor (see Table 5.5). Fish meal and meat and bone meal are rich in calcium due to their bone content.

The physiological regulation of the absorption and secretion of calcium was discussed in Chapter 2. There are also several nutritional factors that affect calcium availability:

(1) High calcium: phosphorus ratios in diets such as alfalfa predispose to the formation of insoluble tricalcium phosphate, $Ca_3(PO_4)_2$. Whereas this is unlikely to create a primary calcium deficiency, it could predispose to a phosphorus deficiency.

(2) High fat diets, e.g. protected fats in rations for high-yielding cows, can reduce calcium availability by forming insoluble soaps.

(3) High dietary intake of vitamin D_3 can assist the physiological mechanisms involved in calcium uptake (see Fig. 2.9). Problems associated with deficiency of calcium and other minerals and trace elements will be considered in Chapter 7.

(4) High concentrations of organic anions, especially phytates in cereals, can form complexes which reduce the true availability of calcium in simple-stomached animals like the pig and man. This is not, however, a major problem for ruminants since the rumen microorganisms contain phytases which break down phytates before entry to the small intestine.

Phosphorus is also required primarily to mineralise the skeleton, and about 75% is found in bone. It is, however, the essential element of adenosine triphosphate (ATP), the energy currency of the body. Feedingstuffs of animal origin and protein-rich by-products of oilseeds such as soya and groundnut are reasonably rich in phosphorus (see Table 5.5). Cereal grains and cereal offals (e.g. bran) contain phosphorus mostly as phytates which are of very low availability unless degraded by rumen microorganisms. Grasses and forages are, in general, low in phosphorus and get lower as they mature. The animals most prone to phosphorus deficiency are those such as growing heifers which are getting all or nearly all their nutrients from grass.

Magnesium is in reasonably plentiful supply in most cattle foods. There is a tendency for magnesium to be deficient in young spring grass. Moreover, the availability of magnesium in grass can be reduced:

(1) by high protein (or non-protein N) in grass which leads to high ammonia production in the rumen and the formation of insoluble magnesium-ammonium phosphate $MgNH_4PO_4.6H_2O$.
(2) by the early application of potassium-rich fertilisers. It is important for dairy farmers to delay the application of potash fertilisers until the acute hypomagnesaemia ('grass staggers') season is over.

Potassium is plentiful in all green foods.

Sodium is relatively deficient in most cattle foods. It is especially important for the maintenance of the normal volume of the blood and extracellular fluid. Moreover cattle like the taste of salt (sodium chloride, or NaCl) and it is included in compound foods partly to enhance palatability. Where primary sodium deficiencies occur, usually a long distance from the sea, cattle will travel many miles to salt licks. It is an important nutrient but cheap and easy to provide.

Vitamins and trace elements

Vitamins and trace elements, especially copper, cobalt and selenium, are of extreme importance in cattle nutrition and medicine and will be considered in Chapter 7. Iron, manganese and zinc are essential elements for metabolism but seldom, if ever, likely to cause problems for dairy cows. Iodine deficiencies can occur, usually, as with sodium, far from the sea.

Nutritive value of foods

Having decided the quantity of specific nutrients required to meet the needs of the dairy cow for maintenance, pregnancy and lactation, it is necessary to define the nutritive value of available foods using the same currency. Textbooks of nutrition and publications from the agricultural advisory services of most nations provide comprehensive tables of nutritive value which include DM, the apparent digestibility of organic matter in DM (DOMD or 'D' value), ME, CP and (at least) the major minerals calcium and phosphorus. Table 5.5 is a relatively short compilation of nutritive values of some common constituents of dry concentrate foods for cattle and Table 5.6 describes bulk feeds. In each

Table 5.5 Typical nutritive values for some common constituents of dry concentrate foods for cattle. All materials are assumed to have a dry matter concentration of 900 g/kg

	ME (MJ)	FME (MJ)	CP (g)	ERDP† (g)	DUP† (g)	Ca (g)	P (g)
	Nutritive value per kg DM						
Cereals							
wheat	13.6	12.9	115	95	8	0.3	0.4
barley	12.8	12.3	114	89	14	0.4	0.4
oats	12.1	10.7	105	82	5	1.0	3.0
Oil cakes and meals							
cottonseed meal (dec.)	12.5	11.2	486	301	133	1.9	12.4
groundnut meal (dec.)	13.7	11.3	570	386	131	1.6	6.3
soyabean meal	13.3	12.7	497	315	145	4.5	7.6
rapeseed meal	12.0	10.8	400	288	57	4.2	8.3
linseed cake	13.4	10.4	334	177	118	4.0	8.0
Miscellaneous by-products							
brewers' grains	11.7	9.0	249	91	120	3.3	4.1
maize distillers' grains	14.7	10.9	317	159	50	3.0	3.8
maize gluten feed	12.7	11.5	207	141	31	0.4	3.4
maize gluten meal	17.5	16.4	666	244	358	1.6	5.0
sugar beet pulp (mol)	12.5	12.3	103	49	38	10.0	3.0
Fish meal	14.2	12.0	686	300	309	80	44
Dairy cake*	13.0	12.1	180	108	50	8.0	6.0

dec = decorticated, mol = molassed.
† ERDP and DUP are calculated at a rumen outflow rate of 0.05 per hour.
* an arbitrary value included for use in ration formulation exercises; resemblance to any existing cake is coincidental.

case nutritive value is described not only in terms of conventional nutrients ME, CP, calcium and phosphorus but also includes best estimates of FME, ERDP and DUP, the latter two values calculated after adjusting SDP values to a rumen outflow rate of 0.05 per hour. Most of these values have been drawn from the 1993 Report of the Interdepartmental Working Party on Protein Requirements for Ruminants.

All these values should be treated with caution. Many of the cereals, oilseed cakes and meals are reasonably constant in composition and so are unlikely to differ from published average values in terms of the crude chemical constituents, crude protein, oil and fibre that define

Table 5.6   Typical nutritive values of some common bulk foods for cattle

	Nutritive value per kg DM							
	DM (g/kg)	ME (MJ)	FME (MJ)	CP (g)	ERDP† (g)	DUP† (g)	Ca (g)	P (g)
Fresh grass 75–80 D	200	12.3	11.4	156	116	22	6.5	3.0
60–65 D	200	9.8	9.1	98	59	26	6.0	2.5
Grass silage, excellent	280	11.4	8.9	175	115	20	6.5	3.0
moderate	240	10.4	7.9	160	105	18	6.0	2.5
Big bale silage	280	9.9	7.9	109	71	20	5.5	2.0
Maize silage	240	11.2	9.0	86	55	5	4.0	3.0
Grass hay	850	8.5	7.3	85	45	20	4.0	2.0
Alfalfa hay	850	9.5	8.3	190	112	25	17.0	3.0
Barley straw, spring	860	7.0	6.5	36	21	0	3.5	1.0
Kale, marrowstem	140	11.0	10.2	160	118	22	21.4	3.0
Swedish turnip	110	13.0	12.4	110	81	10	2.0	1.5
Fodder beet	180	12.6	11.9	60	45	5	1.5	1.5

D = percent digestible organic matter in DM.
† ERDP and DUP calculated at a rumen outflow rate of 0.05 per hour.

simple (or 'straight') and compound foods for legal purposes and are stated on the outside of the bag. The nutritive value of some by-products such as maize gluten feed and distillers' grains varies according to the way they have been processed. Grasses and forages such as hay and silage differ enormously in composition according to such things as stage of maturity and method of conservation to the extent where it is almost meaningless to rely on standard tables.

The variation that exists within classes of food in crude protein, oil (ether extract) and crude fibre is easy to measure in a consistent way, i.e. the same sample sent to different laboratories will yield much the same answer. It is, however, much more difficult to generate consistent and reliable estimates of the more useful descriptions of nutritive value, namely ME, FME, ERDP, DUP. The values in Table 5.5 therefore must be treated as approximate and those in Table 5.6 as very approximate. In practice, the nutritive value of (say) grass silage used in a winter ration for dairy cattle will be individually assessed from chemical or physical analysis. I shall discuss factors responsible for variation in the nutritive value of the more common foods in the next chapter. At this stage, however, I repeat, all descriptions of nutritive value are uncertain. If milk yield (which *can* be measured with certainty) differs from that predicted, it may not be the cow that is at fault but our own estimate of nutrient supply.

Ration formulation

Prediction of appetite

Although DM intake of the individual cow is the greatest single area of uncertainty in formulating rations to meet nutrient allowances, we have to start somewhere. DM intake is assumed, very simply, to be a function of body size (or $W^{0.75}$) and nutrient demand (or hunger) which is itself determined by milk yield. MAFF recommend:

$$\text{DMI (kg/d)} = 0.025W(\text{kg}) + 0.1Y(\text{kg})$$

where Y is milk yield (kg/day).

Thus a 600 kg Friesian cow with a milk yield of 25 kg/day would be predicted to eat $(0.025 \times 600) + (0.1 \times 25) = 17.5$ kg DM/day. This use of body weight (W,kg) rather than metabolic body size $W^{0.75}$ makes this equation very unreliable for small cows like Jerseys and hopeless for an animal as small as a goat.

For those who like Jerseys and linear equations, I suggest:

$$\text{DMI (Jerseys)} = 0.033W + 0.1Y$$

It is, however, better to use $W^{0.75}$. The review by ARC (1980) of DM intake in cows suggests an average value of 135 g DM/kg $W^{0.75}$ per day for cows giving modest amounts of milk and I have based my estimates of DM intake in Tables 5.1 and 5.3 largely on this. More detailed predictions incorporating adjustments for yield and stage of lactation are given in Table 5.7. The problem of individual variation in DM intake is not as great as it might appear since cows that sustain higher yields than predicted must also be eating more than predicted since all the nutrients have to come from somewhere.

Assessment of an existing ration

Consider a group of autumn-calved Friesian cows weighing approximately 600 kg and giving an average yield of 27 kg/day. They are receiving moderate silage (M/D = 10.4 MJ/kg DM) *ad libitum* and

Table 5.7 Predicted DM intakes of dairy cows

Month of lactation	Milk yield (kg/day)	Breed and body weight (kg) Jersey 350	Friesian 600	Holstein 700
1	15	9	13	15
	25	11	14.5	16.5
	35	–	16	18
3	20	13	17	19
	30	16	18.5	20.5
	40	–	20	22
10	5–15	10	15	17

3 kg hay/day. A dairy cake is being fed in parlour only at 0.4 g/kg milk in excess of 10 kg milk/day. This amounts to 6.8 kg per day. Allowances are taken from Table 5.3, nutritive values of foods from Tables 5.5 and 5.6 and DM intake obtained by interpolation from Table 5.7.

Supply of nutrients is set down as follows:

	DM intake (kg/d)	Nutrients per day (MJ or g) ME	FME	ERDP	DUP	Ca	P
Dairy cake	6.1	79.3	73.8	659	305	49	37
Hay	2.6	22.1	19.0	117	52	10	5
Silage	9.3	96.7	73.5	976	167	56	23
Total	18.0	198.1	166.3	1752	524	115	65

ERDP/FME = 10.5 (<11) thus microbial protein supply limited by ERDP

Metabolizable protein = (0.64 × 1752) + 524 = 1645

	ME	MP	Ca	P
Requirement:	196	1713	97	74
Balance:	+2	–68	+18	–9

At this stage the new Metabolizable Protein System (IDWP, 1993) suggests a 10% safety margin for MP by assuming a 10% reduction in supply (i.e. $0.9 \times 1645 = 1480$) giving a negative balance of 233 g MP/ day. It is too soon to say whether or not this added safety margin is justified. My personal inclination is not to use the safety factor mainly because I believe that the MP system may tend slightly to overestimate ERDP requirement of rumen microbes by not including a contribution to QDN from recycled endogenous urea.

Taking these numbers at face value, therefore, we see that this very typical ration is perfectly balanced for ME at an M/D of 11.0. Calcium and phosphorus are both within acceptable limits. The diet is marginally deficient in ERDP with respect to FME although at 10.5% (for a target of 11.0) this is well within the limits of uncertainty. However, since ERDP is the limiting nutrient, MP from microbial protein is calculated from ERDP supply. If the ratio ERDP:FME had exceeded 11, then FME would have been the first limiting nutrient and MP supply would have become

$$(0.64 \times 11 \times 166.3) + 524 = 1694 \text{ g/d}$$

Least cost ration formulation

The commercial and government-funded agricultural advisory services are now able to use the 'number-crunching' power of computers to formulate least-cost rations for dairy cows that take into account:

(1) the amount of home-produced foods and their predicted nutritive value;
(2) the nutritive value and cost of other simple foods such as brewers' grains that the farmer might consider buying as alternatives to a compound dairy concentrate;
(3) the nutrients required in the compound dairy concentrate to balance all other foods;
(4) the combinations of raw materials that will provide all the nutrients required from the compound dairy concentrate at least cost.

The computer can, of course, incorporate calculations far more complex than I have outlined in this chapter in predicting nutritive value and nutrient allowance. The lengthy calculations are very valuable because they can add precision and since they all take place inside the computer they need not concern the farmer, his advisor, nor (you may be glad to know) my readers.

Listed below is the sort of information that can go into and come out of a programme stored in a portable computer carried onto farm by nutrition advisors to the dairy industry.

A feedplan for dairy cows

Inputs

(1) Average cow weight.
(2) Predicted weight change in early, mid and late lactation.
(3) Average milk yield and composition.
(4) Adjustments to standard predictions of appetite to account for, e.g. 'out-of-parlour' feeding or palatability of silage.
(5) Foods available to farmer and their predicted nutritive value.
(6) Cost of home-produced or locally-purchased foods (£/tonne fresh weight).
(7) Milk price (p/litre).

Outputs

(1) 'Best buy' within available range of compound foods – especially in terms of crude protein concentration.
(2) Raw materials needed to produce 'best buy' food at least cost. For obvious reasons this information is normally restricted to the supplier.
(3) Recommended rations for individual cows according to yield and stage of lactation.
(4) Calculated nutrient balance in terms of ME, ERDP and DUP.
(5) Economic analysis of costs and margins using recommended foods.
(6) Economic value of alternative foods.

Table 5.8 sets out a typical example of a computer print-out based on such information that can be given to, and understood by any dairy farmer. In this example the computer has been asked to recommend a compound food to supplement a ration of silage, fodder beet and maize gluten feed already available on farm for cows in early lactation with milk yields ranging from 22–38 litres/day. The computer has selected from the available range of compound foods 'Milkonomic 20', a 20% crude protein compound, and suggested that intake of

Table 5.8 A specimen computer print-out of feeding recommendations for dairy cows in early lactation (for explanation see text)

Input Rolling annual yield 6300 litres/cow
Breed, Friesian
Milk price: 17.3p/litre

TARGET COW weight 600 kg
yield, 30 kg/d, fat = 4.00%, protein = 3.20%
stage of lactation, EARLY; weight change –0.3 kg/day

Foods available on farm	Ration (kg/d)	Price (£/tonne)	DM (g/kg)	ME (MJ/kg DM)	CP (g/kg DM)
grass silage	(45)*	20	240	10.6	160
fodder beet	11.0	18	180	12.6	60
maize gluten feed	3.0	110	900	12.7	207

Output FEEDING RECOMMENDATIONS

Milk yield (litre/day)

	22	26	30	34	38
Ration (kg/d as fed)	22	26	30	34	38
Silage (estimated intake)	45.2	42.6	40.0	37.4	34.8
Fodder beet	11.0	11.0	11.0	11.0	11.0
Maize gluten feed	3.0	3.0	3.0	3.0	3.0
'Milkonomic' 20	0.5	1.6	3.6	5.7	7.8
Margin over concentrate (p/litre)	16.9	16.3	15.4	14.6	14.0
(p/cow,d)	372	424	461	497	533
Margin over all foods (p/litre)	10.4	11.0	11.0	10.9	10.8
(p/cow,d)	229	286	329	369	411
NUTRIENT BALANCE (%)					
Metabolizable energy	100	100	100	100	100
Metabolizable protein	104	102	102	102	102
ERDP	94	96	98	98	98
DUP	130	122	118	112	106

* Limit to silage available on farm, not daily intake

compound should increase from 0.5 kg/day at a milk yield of 22 litres/day to 7.8 kg/day at 38 litres/day yield. This exactly balances ME supply to requirement, which simply reflects the fact that this is the first constraint on the programme. The balance of MP and ERDP are almost exact but DUP supply tends to be in excess of demand, particularly at low yields. It should be apparent however that this is due to the composition of the foods available on farm rather than the compound.

The computer also prints out predicted gross profit margins over concentrates (in this case the compound food) and all food. This information can be used to set targets for economic performance. Alternatively the farmer and his consultant may now proceed to ask the computer a series of 'what if' questions such as:

(1) What if I doubled the intake of maize gluten feed?
(2) What if the M/D of the silage has been over- or under-estimated?
(3) What if I could obtain a source of wet brewers' grains at £25/ tonne?
(4) What if, next year, I decided to grow maize silage?
(5) Can I devise a ration based entirely on forage, fodder beet and 'straights'; (i.e. dry foods such as those listed in Table 5.5) thereby eliminating altogether the need for a compound food?

In each case the computer would re-assess the adequacy of the ration (in this case in terms of ME, MP, ERDP and DUP) and predict the effect on gross profit margins.

Clearly, computer-based systems of this sort constitute a powerful tool for the feeding of dairy cows at high precision in terms of nutrient supply and at least cost. They do, however, need to be treated with professional caution. Even the most advanced of programmes incorporating new concepts such as MP, FME, ERDP, etc. is subject to considerable uncertainty as to (i) nutritive value of foods, (ii) efficiency of utilisation (k_p, k_l), (iii) the contribution of body weight loss to nutrient supply in early lactation and, *most especially*, (iv) uncertainties as to dry matter intakes (DMI).

The ration proposed in Table 5.8 assumes a DMI ranging from 16.0 kg/day at 22 litres/day yield to 20 kg/day at 38 litres/day. Both of these exceed the predicted intake of Friesian cows in the first month of lactation (Table 5.7) and are just about on the limit by month 3. Sometimes, especially when the nutritive value of forage is poor, the computer may print out a ration which is unrealistic for dry matter intake or even dangerous because it calls for a diet excessively rich in starchy concentrates. The print-out is therefore a valuable servant but should not be allowed to dictate feeding policy. It is up to the farmer with his expert adviser to understand both the potential and the limitations of computer-based ration formulation and interpret this information to devise the most economic and healthy practical feeding system for his cows. These practicalities will be discussed in the next two chapters.

Allowances for growing heifers

Detailed tables of nutrient allowances for growing cattle are provided by ARC (1980), MAFF (1980) and NRC (1978). In practice, there is not the same economic justification for precision in formulating rations for heifers as there is for dairy cows. Moreover the nutrient requirements for growth are much more uncertain than those for lactation which makes elaborate ration formulation rather pointless. Nevertheless farmers (in the UK) aim to calve down their heifers for the first time in the early autumn, ideally at 2 to 2½ years of age. To achieve this it is necessary to grow them at a fairly rapid, controlled rate and this requires careful attention to feeding and management.

Table 5.9 Target weights and daily gains for rearing heifers and concentrate allowances in rations for housed calves

	Targets		Concentrate ration
	Daily gain (kg)	End weight (kg)	(kg/day)
Autumn-born calves to calve at 2 years			
0–6 months	0.55	140	1
6–12 months	0.71	270	nil
12–18 months	0.55	370	1–2
18–24 months	0.77	510	nil
Weight at service (kg)	–	330	
Spring-born calves to calve at 2½ years			
0–6 months	0.55	140	1
6–12 months	0.38	210	1
12–18 months	0.66	330	nil
18–24 months	0.33	390	0.5–1
24–30 months	0.77	530	nil
Weight at service (kg)	–	360	

Table 5.9 lists targets for daily weight gains and end weights over 6-month periods for Friesian heifer calves either autumn-born and reared to calve at 2 years of age or spring-born and reared to calve in the autumn at 2½ years of age. In essence, the objective is to feed as little expensive concentrate food as possible. The table lists the approximate amounts of concentrate food required to supplement the

forage provided for housed calves in the form of silage or hay. For the first six months the concentrate ration should contain 180 g/kg crude protein and be generously supplemented with minerals. Calves over 200 kg require about 14% CP in the concentrate feed if their forage is hay. If they are given silage the CP concentration in the concentrate need not exceed 120 g/kg which makes it possible to feed straight cereals such as barley. The exact quantity of concentrate fed should be adjusted according to the quality of forage assessed simply from the weight gains of the heifers. It should, for example, be perfectly possible to achieve 0.33 kg/day gain in 18-month-old heifers given no concentrate provided the quality of the silage is good.

One general caution at this stage: the more reliance that is placed on home-grown forage and straight cereals the greater the risk of mineral deficiencies. A standard mineral supplement sprinkled onto the forage at about 50 g/cow per day can provide the major minerals. Specific problems of trace element and vitamin deficiencies will be considered in Chapter 7.

6 Foods and Feeding Strategies

In most of the major dairying areas of the world, the bulk of the food eaten by cows is grass, maize or lucerne, fresh or conserved. At best, these are close to being ideal foods for adult cattle, although not quite in perfect balance with respect to essential nutrients, and they may contain too much indigestible fibre to meet the requirements of the highly productive animal within the constraint of appetite. Feeding strategies for dairy cows are therefore usually attempts to solve the simple question: 'To what extent should I *substitute* a concentrate ration for grass or forage in order to meet nutrient requirements?'.

A concentrate ration is, by definition, one which contains a high concentration of nutrients (principally ME) per kg fresh weight. In practice this usually implies a mixture of dry foods consisting largely of cereals, cereal by-products and protein-rich extracts such as soya bean meal or fishmeal fortified by appropriate additions of essential minerals and vitamins.

When the differential between the price of milk and the price of concentrate is high and when every farmer is able to sell as much milk as he can produce, many may find themselves providing up to 70% of the ME requirements of their cows in the form of purchased dairy cake. Buying in such a high proportion of food enables them to keep a large number of animals and so generate as much income as possible to set against fixed costs for labour and equipment.

Since the European Economic Community has imposed quotas on milk production from each farm and a supplementary levy which effectively replaces income with a fine on all milk produced in excess of quota, the simple strategy of maximum output has had to be revised. In the long term, a farmer can respond to the regular, repeated reductions in the amount of milk that he is permitted to sell by purchasing quota but within any one year his milk output is fixed in advance. This may be tough in a strictly financial sense but it makes dairy cow nutrition much more interesting. The simple 'more cake, more milk' approach has to give way to a carefully formulated and regularly reviewed strategy designed to achieve annual quota for milk sales exactly and at maximum profit. Since income is fixed, except for seasonal variations in milk price, increasing gross profit margin becomes therefore very much a matter of reducing feeding costs. In most cases this implies a high dependence on home-grown food, especially grass and forage crops. This chapter considers the many foods available to dairy cattle and discusses how they may be used in practical feeding strategies to maximize the biological and economic efficiency of nutrient use.

Grass

The economic value of grass to any farmer is determined by the yield of dry matter and the nutritive value of that dry matter. Although this

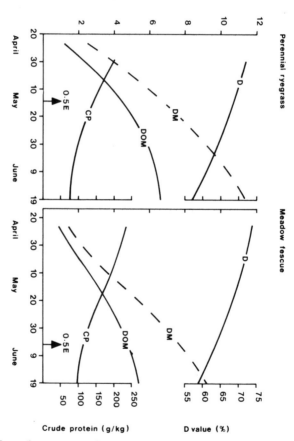

Fig. 6.1 Effect of season on yield and nutritive value of perennial ryegrass and meadow fescue. DM = dry matter, DOM = digestible organic matter, D = digestibility of organic matter (%), CP = crude protein, 0.5E = 0.5 emergence of seed heads.

is a book on cows, not grass, it is useful to examine how yield and nutritive value change as the grass crop grows and matures. Figure 6.1 illustrates two species – perennial ryegrass, a high-yielding, early maturing grass, and meadow fescue, a lower yielding, later maturing species.

Given reasonable growing conditions in the UK perennial ryegrass outperforms other grass species, yielding about 11 tonnes DM/ha by mid-June which is the preferred time for the hay harvest. However, the proportion of fibre has increased steadily and this has reduced the digestibility of the grass, conventionally defined by D value, which describes the *in vitro* digestibility of organic matter in the dry matter (DOMD). In Fig. 6.1 ryegrass cut for hay in mid-June yields 11 tonnes DM at a D value of 60%, giving 6.6 tonnes of digestible organic matter (DOM) with a crude protein concentration of 80 g/kg. Grazed, or cut for silage on 20 May, it yields 5 tonnes DOM at a D value of 67% and CP concentration of 130 g/kg.

Meadow fescue is much later maturing, as assessed by the time taken for 50% emergence of the 'ears' or seed heads (0.5E). It has a higher CP concentration than perennial ryegrass, but a much lower yield. Moreover, despite its later maturity, its D value declines with the advance of the season in similar fashion to that for ryegrass. Fescues and other grass species such as cocksfoot and timothy can usefully be

Table 6.1 Predicted metabolizable energy (MJ/kg DM) of grass from D value and crude protein concentration.

D value	Metabolizable energy (MJ/kg DM) at crude protein (g/kg DM)		
(%)	100	150	200
60	9.5	10.0	–
65	10.2	10.7	11.2
70	10.9	11.4	11.9
75	–	12.1	12.6

incorporated into grazing pastures to provide a more even supply of high-quality grass well into the summer months (Holmes, 1989), but Fig. 6.1 makes it easy to understand why perennial ryegrass is the preferred species for grass crops conserved as hay and especially silage.

The relationship between D value and ME concentration (MJ/kg

DM) is given in Table 6.1 which is derived using the following equation (Holmes, 1980):

$$ME \ (MJ/kg \ DM) = 0.138 \ D(\%) + 0.01 \ CP \ (g/kg) + 0.231$$

A grass with a D value of 65% and CP of 150 g/kg has an estimated M/D of 10.7 MJ/kg DM. A silage crop having this nutritive value would be an excellent winter forage for high yielding dairy cows.

Grazing

Whereas, if the weather is kind, it is possible to harvest grass for silage at the optimum time with respect to yield and nutritive value, it is, of course, not possible to manage a programme of grazing with the same precision. When cows are first turned out onto spring grass there is almost invariably an increase in milk yield (Fig. 3.6). Young, rapidly growing spring grass about 10 cm in height has a D value of about 75% and thus an M/D in excess of 12 MJ/kg DM and a CP concentration of 160–180 g/kg. In terms of the major nutrients it can, in theory, sustain maintenance plus about 30 l/day of milk production. This may not be achieved in practice because, prior to turnout, dairy cows have usually been eating a high proportion of starchy concentrate food and the ruminal microorganisms take time to adapt to the change of diet. Young spring grass is, inevitably, low in fibre. This may reduce the yield of acetate from fermentation and so reduce the concentration of milk fat. Young, highly fertilised grass also tends to be low in magnesium and may predispose to acute hypomagnesaemia or 'grass staggers' (Chapter 7). The most important supplements for spring grass are therefore digestible fibre and magnesium.

Spring and summer grazing should be managed to keep the best possible balance between growth and intake of grass. Overgrazing occurs when the average length of grass that the cows are prepared to eat is less than 10 cm. This does not necessarily imply an absolute shortage of grass, particularly if stocking density is low, but when the length of the grass is short, cows cannot take in enough nutrients in the time available. Obviously as the season progresses an increasing proportion of the pasture becomes soiled by faeces and while this returns nutrients to the soil it is of little immediate use since cows reject regrowths of grass around dung pats. By late summer grazing meadows are dotted with patches of lush-looking long, green grass that cows won't eat. If the grass is undergrazed many plants will mature, become stemmy and form seed heads. This reduces the nutritive value of the pasture and discourages further growth of nutritious young leaves. Such overgrown pastures should be 'topped' with a mower.

Even if grazing pastures are well-managed, the nutritive value of grass usually declines in mid-summer, D value falling to about 60% in dry seasons. In autumn D value may return to about 65% and the grass may look as good as spring grass. However, the capacity of autumn grass to provide nutrients for lactation is poorer than that of spring grass (even at the same D level). It is realistic to assess its M/D at 9.5 to 10 MJ/kg DM.

Grazing strategies are, in practice, a compromise between the need to get the most out of grass and the exigencies of time and expense.

STRIP GRAZING

This is the method whereby each day (or between each milking) cows are given access to a new strip of fresh grass, the access being controlled by moving an electric fence down the line of the field. This method is particularly effective for the first growth of high quality spring grass since very little is wasted as the cows mostly defecate on areas of the field that have previously been grazed down. In late summer and autumn when the pasture is variable in quality and generally poorer, strip grazing is less likely to be worth the effort.

ROTATIONAL GRAZING

This means, in essence, moving cows from pasture to pasture to give the grass time to recover (and to control parasitic roundworms). Cows are kept in paddocks for, say, 2 weeks, or more precisely until the availability of grass is reduced to the point below which intake will decline, and then moved on. Pastures may be rested for 4–8 weeks according to the rate of regrowth.

SET STOCKING

This means stocking a pasture with sufficient cows to ensure that the rate at which they eat the grass exactly matches its rate of growth. Stocking rate in mid-summer would therefore be only about 60% of that during the period of maximum growth in the spring. However, because of defecation and selective grazing, a pasture that was continuously set-stocked by cows all summer would become a complete mess.

In practice, most farmers operate a grazing policy that owes a little to each of these methods, for example, strip grazing the young spring grass and then adopting a compromise between set stocking and rotational grazing during the summer.

It is important to keep a close check on the efficiency of utilisation of the total grass crop both grazed and conserved. The contribution of grass to milk production can be calculated in terms of utilised metabolizable energy (UME) per cow or per hectare per year. The units of UME are usually expressed as Gigajoules (GJ = MJ × 10^3). An example of the calculation of UME from grazed and conserved grass is given in Table 6.2. The first step is to calculate animal requirement for ME per cow, then subtract ME from dairy cake and other purchased foods to estimate UME from grass. In this example the average cow has obtained 41.8 GJ ME per annum from grass, or 78% of its total requirement. A cow fed 2 tonnes of concentrate per annum would only obtain 56% of its total ME requirement from grass. It is, of course, possible to subdivide UME from grass into that from grazing and that from forage. Suppose the average cow in Table 6.2 consumed 2.2 tonnes of silage DM having an M/D of 10.4 from October to April; this would amount to 22.9 GJ ME from silage and 18.9 GJ from summer grazing.

The attraction of the UME calculation is that the farmer can assess

Table 6.2 Calculation of utilised metabolizable energy (UME, GJ/year from grass)

Average production figures

Cow weight = 600 kg, yield = 5500 kg milk plus one 50 kg calf
Dairy cake eaten = 1 tonne, grassland = 0.5 ha/cow

Requirement for ME/year	GJ(MJ×10^3)
Maintenance + pregnancy	26.2
Lactation (5500 kg at 5 MJ/kg)	27.5
Total ME requirement	53.7

ME supply/year

Dairy cake (900 kg DM at 13.2 MJ/kg DM)	11.9
ME from grass (grazed and conserved) per cow	41.8
Utilised ME per hectare	83.6

the performance of his cows against averages and targets for other farms in the area in any one year and also evaluate the consequences of any changes in feeding or grazing strategy.

Legumes

Alfalfa or lucerne (*Medicago sativa*) is a high-yielding, highly nutritious legume which is probably the most important green food for cattle in dry, tropical and sub-tropical environments. It can be grazed but is more commonly conserved or cut and fed fresh to cattle at the optimal stage of growth. In these ideal circumstances it has an ME concentration of about 9.5 MJ/kg DM and a crude protein of 190 g/kg DM. Lucerne is also rich in calcium and magnesium. Most of the nutrients are contained in the leaf. The stem is fibrous and indigestible which means that D value falls sharply as the plant matures. Moreover, lucerne hay, fed in racks, tends to shatter; the nutritious leaves fall out and get lost and the cattle are left with the highly fibrous stems. Lucerne hay is also particularly susceptible to moulding.

Sainfoin (*Onobrychis sativa*) is rather similar in nutritive value to lucerne when young but, once again, its nutritive value declines rapidly as it becomes more stemmy.

The commonest legumes in temperate pastures are the red and white clovers (*Trifolium* species). The bacteria in their root nodules enable them to fix nitrogen and they are drought resistant. In pure swards they have a slightly higher nutrititive value than grass but if eaten in large amounts are liable to cause bloat (Chapter 7), probably due to saponins which cause frothing of the rumen contents by trapping the gases formed during rumen fermentation in small bubbles which cannot be belched.

Grass silage

Silage is, in essence, pickled grass (or alfalfa, maize, etc.). If weather permits, the crop is cut and harvested at a stage of maturity that achieves the best possible balance between yield and nutritive value and packed, consolidated and sealed into a silo to exclude as much air as possible. In these anaerobic conditions fermentation of sugars in the grass, by naturally-occurring microorganisms attached to the grass, forms organic acids which reduce pH to the stage where the crop is preserved against further microbial attack. In ideal circumstances of warm, dry weather when the grass is rich in sugars and the predominant natural organisms are lactobacilli, pH falls rapidly to about 4.0–4.2, bacterial growth ceases and the crop becomes stable after minimal loss of original nutrients. If this rapid fall in pH does not take place, because of a lack of fermentable sugars in the grass, or a relative lack of suitable microorganisms (lactobacilli), or if the crop is exposed to air, secondary fermentation by other microorganisms such as clostridia can occur. The greater the amount of secondary fermentation, the greater the loss of organic matter, the poorer the quality of the organic matter that remains and the closer the silage gets to becoming compost.

The nutritive value of silage is therefore an extremely variable commodity being determined not only by the species of grass and its maturity at the time of harvest but also by the chemical changes that take place during harvest and conservation. Ideally, of course, these chemical changes should be restricted so that the composition of silage is as close as possible to that of the grass from which it was made. However, even the best made silage is a less balanced diet than its parent grass since some protein will have been degraded (to QDN) and, more critically, the fermentable sugars in grass (QFE) will have been converted to organic acids (lactic acid, acetic acid, etc.) which can be utilised as an energy source by the tissues of the cow but not by the microorganisms in the rumen. Conservation of grass as silage therefore inevitably reduces both the rate and extent of fermentation of organic matter, unbalances the supply of fermentable energy and degradable protein, and so reduces the yield of microbial crude protein to the abomasum.

Because so many factors can affect the nutritive value of silage, it is not something that can be assumed from inspection or from published tables. Core samples should be taken from each clamp and sent to a laboratory for analysis. Those wishing to know a great deal more should consult *The Biochemistry of Silage* by McDonald, Henderson and Heron (1991). I shall consider, very briefly indeed, interpretation of the results of silage analyses.

Dry matter

The dry matter (DM) concentration in grass at the time of cutting for silage is unlikely to exceed 20%. The nutritive value and palatability of silage from such grass is likely to be maximal at a DM concentration of 25%. Very wet silages (below 20% DM) give rise to large quantities of effluent and are difficult to preserve without the use of additives. Very dry silages (over 40% DM) tend to have a high pH (over 5.0) and may suffer from spoilage due to secondary fermentation if a large area of the silage face is left exposed. Wilting of grass before ensiling is less popular than it used to be. While it may improve conditions for a good fermentation by concentrating sugars it rarely seems to improve the palatability of silage from young grass cut on a warm day in the spring and usually reduces digestibility by 2–3 percentage units. It may be desirable (although difficult) to wilt wet autumn grass with a low sugar concentration but this would be hardly suitable for lactating cows anyway. DM concentration can be assessed on farm simply by squeezing the silage. If moisture cannot be squeezed out, DM is over 25%; if a tight squeeze leaves a little moisture on your hands, it is between 20 and 25%; if it runs through your fingers it is below 20%.

Effluent production from silage is negligible at DM concentrations above 25%, rising to about 50 litres per tonne of ensiled grass at 20% DM and 200 l/tonne at 15%. Wet silage is not inherently unpalatable, i.e. the ingested water is rapidly absorbed and does not constitute a problem of gut fill. However, wet silage made without additives can be bad silage both in terms of palatability and nutritive value through deterioration of organic matter. Finally, the wetter the silage, the wetter the faeces. There can be no doubt that the foot problems of British dairy cows kept over winter in cubicle houses can be attributed, at least in part, to being compelled to stand in wet, corrosive slurry (Chapter 8). Conservation of grass as wet silage rather than dry hay has been a conspicuous improvement, measured strictly in terms of nutrient supply. It has, however, generated new problems, particularly those of effluent and slurry.

Metabolizable energy

For many years most estimates of the ME concentration of silage were based on measurements of fibre content. In England the approved method was the modified acid-detergent or MAD fibre technique. Advisory services in Scotland were more likely to measure the digestibility of organic matter *in vitro* by a two-stage process involving incubation with rumen microorganisms followed by digestion in acid and pepsin. The MAD fibre technique was relatively cheap but very inaccurate, the *in vitro* digestion technique much more accurate but slow and expensive.

Much time and effort has been invested to develop improved techniques for predicting the ME concentration (M/D, MJ/kg DM) of silages on a commercially viable basis. At present the method of choice is based on near infra-red (NIR) spectroscopy. Bombardment of the sample by NIR irradiation generates a complex wave form which reflects both its physical and chemical composition. Very small variations between samples can be analysed by computer and correlated with direct *in vivo* measurements of M/D obtained from digestibility trials. It is then possible to generate complex equations relating M/D to variation in (say) six peaks in the NIR spectrum. These predictions are a great improvement on those based on a single measurement (such as MAD fibre) but it must be stressed that they are entirely empirical and can only be used to predict M/D (or other criteria of nutritive value) by reference to known values for standards of similar composition to the test materials, i.e. a prediction equation based on NIR values obtained for ryegrass silages could not be extrapolated to maize silage.

From time to time the advisory services revise their equations to predict M/D in grass silages, usually upwards. In 1980 a typical M/D value for grass silage would be 10.0, in 1990, perhaps 11.0. More than half this apparent increase can be attributed to revision of the prediction equations rather than to any real improvements in silage making. There is general agreement that the M/D value of silage was underestimated in the past and feeding trials with lactating cows fed excellent silage are consistent with M/D values of 11.0 or higher. However, some silage analyses based on NIR spectroscopy are now predicting M/D values as high as 12.0. I think it very unlikely that M/D could exceed about 11.4 for high-yielding dairy cows consuming large quantities of silage *ad libitum* and with a fractional outflow rate of dry matter from the rumen of 6% or more per hour.

The other factor that should, ideally, be taken into account when estimating the energy value of silage is the proportion that exists in the form of prefermented acids. On average they may constitute 10% of ME but, in some cases, exceed 20%. The ratio FME:ME for the grass silages in Table 5.6 is 75–78%. When silage is fed alone the ratio of microbial crude protein yield to ME (as VFAs) is only about 70% of that from a diet properly balanced with respect to fermentable energy and degradable N (McDonald and others, 1991). One of the essential features of diet formulation for dairy cattle is to overcome the nutritional imbalance created by feeding material (such as silage) in partially prefermented form.

Protein

Even in the best preserved silages about 60% of the crude protein in the grass crop is degraded to some extent into peptides and individual amino acids. In wet, or badly consolidated silages, clostridia and other microorganisms increase the breakdown of proteins and amino acids causing considerable losses of original crude protein as ammonia, an increase in the ratio of QDN to SDN and, in some cases, accumulation of large amounts of potentially toxic nitrogenous compounds such as histamine. Laboratory analyses describe the proportion of total N present as ammonia (NH_3) as an indicator of fermentation quality. As a rule of thumb:

% NH_3–N as % total N	Type of fermentation
less than 5	excellent
5–10	very good
10–15	satisfactory
15–20	poor (slightly butyric)
20 plus	bad (butyric)

In an excellent fermentation where lactobacilli predominate and pH fall has been rapid, the silage has a clean, if somewhat clinging, acid smell. Poorly fermented silages where clostridia and aerobic organisms have become established generate large amounts of ammonia and VFA, especially butyric acid has a characteristic sickly smell. Such silages are not only poor in nutritive value but very unpalatable.

Silage additives

The main types of silage additive are listed in Table 6.3. Most are designed to preserve the crop in the best possible state either by encouraging rapid fermentation with lactobacilli or by inhibiting secondary fermentation by spoilage organisms such as clostridia. The rate of fermentation can be stimulated by ensiling the grass with a source of quickly fermentable energy. Sugar sources such as molasses (or molassed sugar beet pulp) are particularly effective because

Table 6.3 Types of additive for silages made from grass or whole-crop cereals

Mode of action	Class	Examples
Fermentation stimulants	carbohydrate source bacteria enzymes	molasses, cereals, whey, beet pulp, lactobacilli, cellulases and hemicellulases
Fermentation inhibitors	acids 'sterilisers'	formic acid, sulphuric acid, formaldehyde, sodium hydroxide, sodium nitrite
Nutrients	energy source NPN source	beet pulp, cereals, urea, biuret
Absorbents	nutrients non-nutrients	straw, beet pulp, polymers, bentonite

lactobacilli are sugar fermentors. Whey also works but is very wet. Starchy cereals are somewhat less effective. Silage inoculants containing bacteria (lactobacilli) or enzymes (cellulose and hemicellulose) can accelerate fermentation to low pH but only if they have sufficient QFE to work on. A short-chopped, properly harvested, well consolidated silage made in good weather at the right stage of maturity needs no additive to ensure good preservation and palatability. It is therefore fair to argue that inoculants work best on those silages that need them least.

When the crop is damp, cold, lacking in sugars or poorly chopped a good natural fermentation cannot be guaranteed. Inoculants *plus sugars* can improve matters but in these circumstances it is usually better to inhibit fermentation altogether either by the use of strong acids (formic or sulphuric) or with a steriliser such as formaldehyde or sodium nitrite. Acids are corrosive and difficult to handle but very effective. Sterilants are safer to work with but probably slightly less effective.

Material may be added to grass at the time of ensiling either to absorb water (and so reduce effluent) or in an attempt to redress the energy:protein imbalance in silage (too little QFE, too much QDP). Absorbents having no nutritive value such as the colloidal clay, bentonite, certainly reduce effluent but tend also to reduce nutritive value. So far as the cow is concerned, molassed sugar beet pulp is ideal since it absorbs moisture, contributes QFE to encourage rapid fermentation and contributes a rich source of digestible energy (SFE) in its own right when the silage is fed. However, it is relatively expensive, laborious to spread among the grass in the silo and produced at the wrong time of year. Straw has also been used as an absorbent/nutrient. As an absorbent it is effective but it significantly reduces nutritive value and is not recommended for silage intended for dairy cows. The strongest argument against mixing other sources of nutrients with grass at the time of ensiling is that some material will inevitably be lost during fermentation. If cereals, or sugar beet pulp, or even molasses are to be fed with silage it makes more sense to mix them in at the time of feeding.

Hay

As indicated in Table 5.6, hay has a poorer nutritive value than silage simply because it is cut when the grass is more mature and fibrous. Typical values are $M/D = 8.5$ MJ, $CP = 85$ g/kg DM. Whereas green silage retains most of the original carotene (a precursor of vitamin A), this is lost during sun-drying of hay. However, the concentration of vitamin D is increased. Since the high-yielding dairy cow requires an M/D of about 11.0 in her overall diet, it becomes almost impossible to devise a balanced diet based on hay with an M/D of 8.5 as the major source of forage, which is why it has been largely superseded by silage in the feeding of dairy cows. It is, however, palatable, relatively easy to handle, a good source of digestible fibre and remains, at best, a splendid contributor to the diet of sick cows, young calves and growing heifers, or cows with low butterfat concentration – perhaps because they are at spring grass. Badly made, mouldy hay is not only nutritionally destitute but unpalatable and potentially dangerous to stock and stockmen.

Straw

Untreated straws have no significant role to play in the feeding of dairy cows and are best thought of as bedding. In recent years however, techniques have evolved for treating straws with chemicals to disrupt the lignified structure of the cell walls and so render more of the cellulose available for fermentation. There are two main ways of making nutritionally improved straw. The first is to use sodium hydroxide (NaOH) which can be applied in liquid or solid form to straw. This increases M/D to about 8.5 making it comparable in energy terms to moderate hay. The alternative, more difficult method is to create an airtight seal around stacks of straw and admit gaseous ammonium hydroxide (NH_4OH). The advantage of this method is that it increases not only M/D but also non-protein N concentration to the point where it is just about sufficient to support optimal microbial synthesis.

The production of nutritionally-improved straw (NIS), using NaOH if not NH_4OH, has made a significant contribution to ruminant nutrition. It can be made sufficiently cheaply to justify inclusion, at small concentrations, in least-cost formulations for compound foods. Moreover, it retains a considerable amount of sodium in the form of sodium carbonate and bicarbonate and these are excellent buffers against excess acid production in the rumen when cattle are fed a diet that is very rich in starch.

Ensiled whole-crop cereals

Maize or corn silage is the forage crop of choice for dairy cattle wherever it can be grown and harvested reliably and well. It is good as or better than the best of grass silages as a source of energy (Table 5.6). Considerable starch is retained so it is well balanced between QFE and SFE. It lacks ERDP but this can be provided at low cost by non-protein nitrogen sources such as urea. It normally ferments well without additives to a low pH (usually below 4.0). It is possible that this is slightly too acid both in terms of palatability and optimal rumen fermentation. Maize silage certainly stings if it comes in contact with a cut finger. In the UK, dairy cows have been shown to perform extremely well when their forage ration was a 50:50 mixture of grass and maize silage, which is as expected since this achieves an excellent balance between the supply of fermentable energy and degradable nitrogen.

In recent years there has been an increasing interest in the production of forages for dairy cattle based on whole-crop wheat or barley, treated with urea which hydrolyses to ammonia and preserves the crop at a pH of about 8.5. However, there have been problems of aerobic deterioration and conservation methods are still being improved. Ammonia concentrations in the crop as fed are high

(300 g/kg total N) which makes it pretty pungent. In trials conducted at Wye College by David Leaver, whole-crop wheat gave disappointing results when fed as the sole forage to dairy cows and milk yields improved (slightly) as progressively more (moderate quality) grass silage was added to the mixture. Current wisdom would suggest that whole-crop wheat or barley silage should therefore be fed in combination with grass silage. Pending newer and better particulars, I remain unconvinced by the merits of this crop. Grass silage, for all its faults, appears to be the better crop where grass grows well and I cannot think of any good reason for ensiling whole-crop wheat on farms that can grow maize.

Other green foods

Other than grasses, the most popular green foods for cattle are the brassicas such as kale, rape, cabbage and the tops of root crops such as turnips and swedes. The leaves and stems of kale and rape are rich in the major nutrients, containing about 11 MJ ME and 160 g/kg CP in dry matter. They are rather unbalanced with respect to calcium and phosphorus (see Table 5.6) and contain a number of toxic substances which can, in excess, cause a deficiency of thyroid hormones and/or haemolytic anaemia. Both these problems can be avoided by restricting the intake of kale or rape to a maximum of one-third of total DM intake.

Swede and turnip tops are probably as safe as kale or rape, but mangolds, sugar- or fodder-beet tops can provoke severe scouring unless wilted before feeding.

Root crops

Root crops fed whole, or chopped, to cattle include turnips, mangolds and fodder beet. Although they differ in water content from about 82% for fodder beet to 90% for turnips, the nutritive value of their dry matter is rather similar (M/D 12.6, CP 60–100 g/kg, Table 5.6). In all cases over 50% of DM is in the form of sugars which are not only highly palatable but provide an excellent source of quickly fermentable energy. This makes them particularly suitable for feeding in chopped form along with grass silage which is unbalanced with respect to rumen degradable N and fermentable energy (see Fig. 2.7). At the University of Bristol farm we routinely include about 10 kg/cow of chopped fodder beet with the daily ration of grass silage for our dairy cows. In countries such as Denmark cows may be given as much as 40 kg fodder beet per day. In these circumstances it is particularly important to ensure that the roots are as free as possible from soil contamination and it may be necessary to hose them down before feeding.

Cereals

The nutritive values of the major dry cereals, wheat, barley, oats and maize, were given in Table 5.5. They are all rich in starch which is an excellent, safe source of ME for cows provided that it is properly incorporated into a balanced diet. It is important to roll or otherwise crush the husk of cereals fed to cattle (but not sheep) since a proportion of whole grains pass rapidly through the reticulo-omasal orifice and appear unchanged in the faeces.

Oats are traditionally considered to be the safest of the cereals when fed 'straight'. This is because they contain the highest ratio of slowly fermentable fibre (and unfermentable oil) to rapidly fermentable starch and so are least likely to provoke an explosive fermentation leading to ruminal acidosis and bloat. Rolled barley and wheat, having relatively less fibre and more starch, are fermented more rapidly and so are more prone to cause problems if cattle are allowed to gorge them.

The starch in wheat, barley and oats is normally fermented totally to volatile fatty acids giving a relatively high proportion of propionate to acetate. The protein in these cereals is about 80% degraded in the rumen and so acts predominantly as a source of microbial protein. The ratio of degradable protein to ME in barley is 6.8 which means that it requires supplementation to provide adequate RDP for microbial protein synthesis.

Uncooked maize starch is not completely fermented in the rumen. Most of the maize starch that escapes fermentation is reduced to monosaccharides by acid digestion and the rest will probably be fermented in the hind gut. Flaked maize is produced by cooking the grain in steam and passing it through rollers. This increases the digestibility of the starch in both the rumen and abomasum and makes it more suitable for young calves.

The starchy cereals are of major importance in the feeding of dairy cows since they are able to increase the M/D of a ration above that which can be provided by forage alone and so help to meet ME requirement within the constraints of DM intake. The feed compounder is not concerned whether (say) barley is or is not better than wheat as a 'straight' food for cows; he considers both principally as sources of starch energy in a mixed ration. If, at any particular time, wheat provides the required nutrients at a lower cost than barley, then wheat will appear in the ration formulation.

If cows consume starchy foods to excess they can get bloat, ruminal and metabolic acidosis, which may, in severe cases, lead to death or chronic crippling diseases such as laminitis. Thus starchy foods should always be fed with caution. If a cow eats a substantial quantity of a starchy concentrate food in a single rapid meal (say 5 kg of cake while in the milking parlour) this will ferment rapidly in the rumen causing a fall in pH, some destruction or inhibition of the cellulose (i.e. grass) fermenting bacteria, and an increased production of propionate relative to acetate. A relative decrease in acetate reduces the butterfat concentration of milk. A reduction in cellulolytic bacteria reduces the rate of digestion of forage, thereby increasing retention time in the rumen and so reducing appetite. Overfeeding starchy concentrates therefore reduces the capacity of the cow to utilise home-produced forages.

As a general rule it pays either to restrict starchy concentrates to a maximum of about 4 kg at a single feed, which may require out-of-parlour feeding, or to ensure that they are well mixed and diluted with more fibrous, slowly digested material, which may require complete-diet feeding.

By-products

In Chapter 1 I illustrated the extent to which the modern dairy cow is fed on by-products of industrial processes primarily designed to extract food for man from a wide variety of plant materials. The feeders of cattle (especially beef cattle) have worked on the assumption that almost any organic matter containing carbohydrate and protein is worth a try and no by-product is too bizarre to escape the attention of the nutritionist, e.g. coffee grounds, kapok cake, guar meal, chicken litter and even recycled cattle faeces and rumen contents. These last three have been used as sources of non-protein N for beef cattle fed predominantly cereal diets on feedlots. In the section below I shall consider briefly only some of the more common, wholesome by-products, paying especial attention to those which are often fed straight rather than incorporated into pelleted compound foods.

Oilseed cakes and meals

The major oilseeds are grown in the tropics or sub-tropics; they include soya bean, groundnut, cottonseed and linseed. In recent years, oilseed rape has become an important crop in the cool temperate zones. The oilseed cakes and meals are the residues left after extraction of the oil and are rich not only in protein but also in ME (see Table 5.5) which makes them excellent supplements for starchy cereals in concentrate rations. The oil is extracted from the seeds either by heat and pressure or by chemical means using an organic solvent such as hexane. Linseed cake, for example, is produced by the 'expellar' process, which employs

Table 6.4 Amino acid composition of some livestock feeds (g/kg DM)

		Limiting amino acids			
		1st	2nd	3rd	
	Crude protein	Lysine	Threonine	Histidine	Methionine + cystine
Milk, dried skim	340	24	14	11	9.6
Fish meal	657	45	26	12	24.6
Soya bean meal	520	30.7	19.5	10.6	16.2
Groundnut meal (dec.)	450	14.4	12.2	10.9	11.6
Maize gluten	250	6.0	9.4	6.6	11.0
Linseed meal (exp.)	340	12.3	12.9	6.1	8.8
Field bean meal	284	18.0	12.2	7.2	5.7
Rapeseed meal	396	19.1	17.8	10.5	8.2
Amino acid proportions (g/kg CP) relative to skim milk					
Fish meal		0.97	0.94	0.56	1.33
Soya bean meal		0.87	0.95	0.66	1.16
Groundnut meal (dec.)		0.45	0.66	0.75	0.91
Maize gluten		0.34	0.91	0.82	1.56
Linseed meal (exp.)		1.51	0.92	0.55	0.91
Field bean meal		0.90	1.04	0.79	0.71
Rapeseed meal		0.68	1.10	0.81	0.71

heat and pressure and leaves a residual oil content in excess of 50 g/kg. Soya bean meal is solvent extracted and contains less than 20 g/kg oil.

SOYA BEAN MEAL

Soya bean meal is the most popular protein-rich vegetable-based by-product in animal feeding. It is rich in CP (over 500 g/kg) and the protein is of high biological value, having an excellent balance of amino acids. Table 6.4 lists the composition of a range of foods in terms of CP and the most important of the essential amino acids. Lysine is defined as the first limiting amino acid which means, in effect, that as the protein concentration of a normal ration for farm animals is decreased so lysine becomes the first amino acid to fall short of tissue requirement. Table 6.4 shows that the balance of amino acids in soya bean meal is not quite as good as in an ideal protein like casein in skimmed milk, but it is reasonably well balanced with respect to lysine (g/kg CP), threonine, methionine and cystine, slightly deficient in histidine, although not to a degree that is likely to be of commercial significance.

One potential problem with conventional soya bean meal is that the protein is highly degradable in the rumen so is not particularly suitable for inclusion in rations calling for substantial quantities of UDP. It is however possible to purchase 'protected soya' which has been exposed to formaldehyde or some other process to reduce degradability. There are a number of toxic, allergenic and antinutrient substances in soya bean meal (as there are in nearly all the other high protein vegetable by-products). Most of these are inactivated by rumen microorganisms. It is however necessary to destroy them, usually by heat treatment, when soya bean meal is fed to simple-stomached animals (which includes young calves) or when it is treated to reduce rumen degradability and so convert it into a source of high quality UDP.

GROUNDNUT MEAL

Groundnut meal (decorticated) is rich in CP but the balance of amino acids is decidedly inferior to that of soya bean meal. The lysine concentration (g/kg CP), for example, is only 45% that of skim milk. This makes groundnut meal unsuitable for simple-stomached animals but, since it is highly degradable, it can be an excellent source of RDP. The greatest problem with groundnut meal is the risk of fungal contamination producing aflatoxins which are extremely poisonous to most simple-stomached species and have been known to cause death in adult cattle.

RAPESEED MEAL

Rapeseed meal has a lower protein content than soya bean meal and is inferior in terms of its proportions of lysine and the sulphur-containing amino acids methionine and cystine. Different strains contain various amounts of toxic and antinutrient substances. Even if these are minimised, large inclusions of rapeseed meal in cattle diets tend to reduce food intake. It is used in dairy nutrition mostly as a relatively minor constituent of commercial pelleted concentrates.

LINSEED CAKE

Linseed cake is highly palatable and, being an expellar product, is relatively rich in oil. It is believed, with good cause, to put a 'bloom' on cattle destined for the show ring. Moreover the oil that is digested in the duodenum has the effect of elevating M/D. Its amino acid composition is, once again, inferior to soya bean meal (Table 6.4).

FIELD BEAN MEAL

The field bean (*Vicia faba*) has been grown for many years in the UK and fed to cattle, either whole or as a meal. It is a crop that has always impressed the nutritionist because of its excellent amino acid composition (Table 6.4) but has proved less impressive to the farmer since it is a dismal crop to harvest at a dismal time of the year.

Foods of animal origin

Although it is not 'natural' to include foods of animal origin in the diet of ruminants, cows need nutrients such as essential amino acids, calcium, phosphorus, etc. in much the same way as omnivores like ourselves or even a true carnivore like the tiger. If these essential nutrients can be provided in wholesome form from foods of animal origin surplus to our own needs (i.e., by-products) then it is, to my mind, sensible and thrifty to ensure that such food is not wasted but fed to animals. Fishmeal can be a highly valuable constituent of diets for dairy cows as a source of high quality protein, well balanced with respect to amino acids, calcium, phosphorus and most of the vitamins.

Moreover, at least 50% of animal protein escapes degradation in the rumen which ensures an excellent balance of amino acids absorbed from the small intestine. The most important product in cattle nutrition is white fish meal which contains, by definition, not more than 6% oil. Until recent years white fish meal had been compared with vegetable products such as soya bean meal simply in terms of the quantity and quality of protein supplied at unit cost and was usually found to be relatively too expensive to merit incorporation to any great extent in dairy rations. However, recent trials with white fish meal suggest that it may have a synergistic effect on other elements of the diet, especially grass silage. The explanation for this is not entirely clear but it appears that fish meal enhances microbial growth in the rumen and so improves both the rate and extent of fermentation of silage. Not only may this increase the supply of ME and MP from a fixed quantity of silage, it may also reduce rumen fill and so encourage an increase in DM intake. This, in turn, would reduce the M/D concentration required to meet ME requirement, thus reduce the inhibition of cellulose digestion caused by eating starch, and so further improve the utilisation of forage . . . and so on.

Table 6.5 presents results of a trial conducted at the ICI Research Station at Jealott's Hill (UK) to examine effects of feeding relatively small quantities of concentrates containing a high proportion of fish protein to cows getting very high quality silage (M/D = 11.3 MJ/kg DM) (Reeve, Baker and Hodson, 1986). The quantities of concentrate

Table 6.5 Milk yield and composition from cows in early lactation given small quantities of protein-rich concentrates as a supplement to high quality silage (M/D = 11.3 MJ/kg DM)

| | Protein concentration in concentrate (g/kg) | | | |
	210		400	
Concentrate ration (kg/d)	3.0	6.0	3.0	6.0
Silage intake (kg DM/d)	13.1	11.1	13.9	12.2
Estimated ME intake (MJ/d)	182	193	191	207
Yield (kg/d)				
milk	26.7	28.1	28.7	28.4
fat	1.05	1.06	1.11	1.12
protein	0.77	0.84	0.85	0.85
Liveweight loss (kg/d)	0.44	0.27	0.29	0.30

From Reeve, Baker and Hodson (1986)

fed during the first 100 days of lactation were only 3 or 5 kg/day with a crude protein concentration of 210 or 400 g/kg. The observed milk yields were good without being spectacular. Table 6.5 does, however, illustrate the point that when the ME content of silage is very high and yields are not in excess of 30 kg/day, there is little need to provide extra ME from concentrate but it can be used to provide extra MP. In this example 3 kg/day of concentrate at 400 g/kg CP produced as much milk as 6 kg/day concentrate at either protein level.

Speaking in a strictly nutritional sense, meat and bone meal emerging from abattoirs and knackeries slaughtering cattle, sheep and pigs is similar to fish meal and has made an important contribution to the diet of farm animals. Before 1980 most rendering processes were designed, for sound commercial reasons, to extract as much fat as possible for sale as tallow. This involved prolonged cooking on a batch basis followed by further fat extraction with solvents. After 1980, again for commercial reasons which seemed valid at the time, plants switched to a continuous rendering process and, especially, cut out the process of solvent extraction. Starting in 1986 cows began to exhibit the distressing signs of what has come to be known as bovine spongiform encephalopathy (BSE) or 'mad cow disease'. It is now clear that this was caused by an agent similar or identical to that responsible for scrapie in sheep. It is almost certain that cows were ingesting this agent (whose nature is still unknown) prior to 1980 and there may have been occasional, sporadic, undiagnosed cases of BSE in cows for as long as sheep have had scrapie (i.e. hundreds of years). However the change to a continuous rendering process without solvent extraction, which increased the fat content of meat and bone meal from about 5% to over 10% has increased the infectious dose of the BSE agent to a level sufficient to kill over 40,000 cows in the UK in 1991. I describe this disease in more detail in Chapter 10. Current epidemiological theory is that all (or nearly all) cases of BSE in cows can be attributed to eating infected material and that transmission from cow to calf never occurs (or hardly ever). The implication of this is that the incidence of the disease will decline fairly rapidly starting in 1993 (Wilesmith, Ryan and Atkinson, 1991). Time will tell. We can be sure of one thing. Institutionalised cannibalism (recycling of cows through cows) may make excellent sense within the strict parameters of nutrient supply but it does carry real danger and the new taboo on recycling of meat meal is here to stay.

Miscellaneous by-products

Sugar beet pulp

Dried sugar beet pulp (see Table 5.5) is an excellent, highly palatable source of ME. It contains only a very small amount of residual sugar and the main source of energy is in the form of digestible fibre. This is fermented more slowly than starch so is less likely to upset the rumen microorganisms (and so forage digestion). It also yields a high ratio of acetate to propionate and so encourages the synthesis of butterfat. It is similar to cereals in CP but relatively rich in calcium. Its palatability and stable fermentation make it a particularly suitable third or 'lunch-time' food for dairy cows, which provides an element of novelty and encourages a higher overall DM intake. Magnesium-enriched sugar beet pulp is an excellent in-parlour food for dairy cows on spring grass since it balances the supply of digestible fibre, magnesium and metabolizable protein.

Maize gluten

Maize gluten is a by-product of starch extraction from maize and is available in large quantities. Maize gluten feed is normally sold at 20% crude protein and is an excellent source of ME (M/D = 12.7, Table 5.5) principally in the form of digestible fibre. It has, however, rather been oversold as a source of protein. Prolonged exposure of maize to heat during the process of starch extraction creates Maillard reaction products, complexes of carbohydrate and organic nitrogen in which the original amino acid structure has been destroyed. These Maillard reaction products appear as increased ADIN (acid-detergent insoluble nitrogen: see Chapter 2), some of which may be degradable in the rumen but none of which can contribute to amino acid supply from DUP. As a general rule the darker the sample of maize gluten, and the more 'Marmitey' its taste, the poorer its protein value. The major feed companies are aware of the differences in protein value of different samples of maize gluten feed. This implies that some of the darker samples which are available to the farmer as 'straights' are those which have been rejected by the compounders. I repeat, maize gluten feed is an excellent source of ME but cannot be considered a sufficient and satisfactory source of MP.

Maize gluten meal is sold at 60% crude protein and most of this exists as undamaged true protein. It tends to be relatively deficient in lysine (Table 6.4) but provided this need can be met from elsewhere (e.g. microbial protein) maize gluten meal is an excellent source of DUP. However, because it is good, it is expensive.

Brewers' and distillers' grains

These are the residues after extraction of starch by malting and fermentation from barley. They may be fed wet or dry. In both cases they, like maize gluten, are excellent sources of digestible fibre. However, they usually contain 10% fat and this may interfere with rumen fermentation and restrict intake if they form too large a proportion of the overall diet. Their protein value is extremely variable, depending on the extent to which the protein has been subjected to heat and moisture. Wet brewers' grains have a low ADIN concentration which suggests that the protein is largely undamaged. In some dark distillers' grains well over 50% of crude protein has been converted to ADIN and may be considered unavailable (Webster, 1992).

Wheat bran

Wheat bran is relatively rich in CP though a substantial proportion of it may be in the form of ADIN. It is too low in ME to be a serious contender with other by-products in dairy cow feeding, particularly since it has acquired a cash value out of all proportion to its nutritive value because of its appeal to humans seeking a natural laxative.

Balancing forages and concentrates

So far I have assumed that the DM intake of a dairy cow is a simple function of her size, milk yield and stage of lactation (Table 5.7). This is an obvious oversimplification since the constraint on appetite imposed by gut fill will be far greater for a slowly fermented material with large particle size like grass silage than for a milled mixture of rapidly fermentable starchy concentrates. Thus a cow given both silage and concentrates to eat will be able to consume more DM than one given silage alone. However, as the intake of concentrate increases, so silage intake declines at an increasing rate. Figure 6.2 illustrates this for a typical Friesian cow fed increasing amounts of concentrate in association with either a high or low quality grass silage (high quality M/D = 10.8, CP = 170 g/kg; low quality M/D = 9.5, CP = 140 g/kg). DM intake (DMI) for the high quality silage fed alone is 14 kg/day.

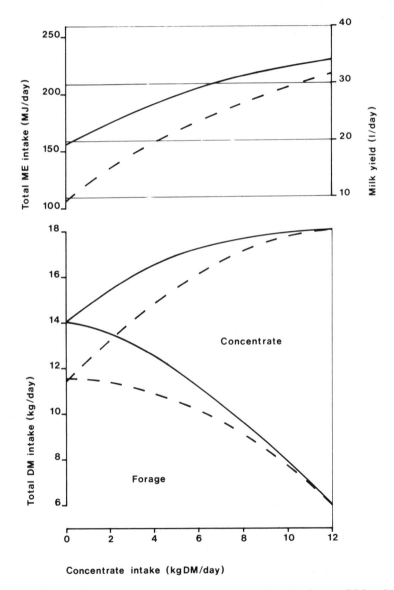

Fig. 6.2 Effect of increasing concentrate intake on total matter intake. DM = dry matter, ME = metabolizable energy. Silage M/D = 11.0 —, 9.5 — — — .

The addition of 4 kg DM of concentrates increases total DMI to 16.5 kg/day, a net gain of 2.5 kg DM. A further increase in concentrate intake from 8 to 12 kg DM/day only increases total DMI from 17.6 to 18.0 kg/day. At this level of concentrate feeding concentrate is almost entirely substituting for, rather than supplementing, forage. With the silage fed alone DMI is only 11.4 kg/day and increasing concentrate has a substantial effect on total DMI up to about 8 kg/day.

Figure 6.2 also illustrates the effect of increasing concentrate intake on total ME intake in relation to ME allowances for cows yielding 10 to 40 litres milk per day. In this example the concentrate is a conventional starchy mixture (M/D 13.2, CP 160 g/kg). With the high quality silage this typical cow consumes enough ME to sustain 30 l milk/day when given 6 kg DM of concentrate. Increasing concentrate consumption to 12 kg DM/day only provides another 22 MJ/day – enough, in theory, to sustain only another 4 litres of milk production. If the only silage on offer is of poor quality then it is probably necessary to feed up to 10 kg of this concentrate per day to sustain yields of

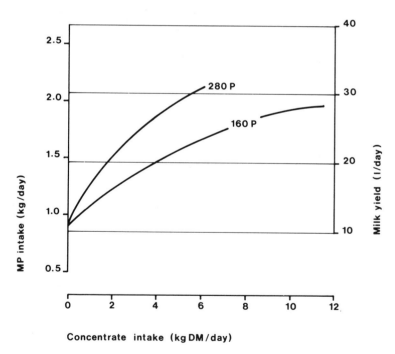

Fig. 6.3 Effect of increasing concentrate intake on the supply of metabolizable protein (MP) in relation to requirement for lactation for concentrates with 280 and 160 g/kg crude protein (P).

30 l/day or over. It should be clear that when the ME value of forage is high, increases in the concentrate ration rapidly reach the point where increments of ME intake (and indirectly milk yield) become very small indeed. Since concentrates are about twice as expensive as grass silage per unit of ME, gross profit margins will fall.

A cow given high quality silage may need only 6 kg concentrate to achieve maximum ME intake. However, many dairy farmers, rightly or wrongly, look on concentrates largely as a supplier of protein. Figure 6.3 indicates the supply of MP when a cow is given high quality silage supplemented by either (a) a conventional cereal-based concentrate mixture with a CP of 160 g/kg, or (b) a concentrate containing 280 g/kg high quality protein with a degradability of 0.6. In this example 6 kg of high protein concentrate supplies as much MP as 12 kg of a conventional cereal-based concentrate. Since there is no significant improvement to ME intake by increasing the concentrate ration above 6 kg/day (Fig. 6.2) it follows that 6 kg of a 280 g/kg CP concentrate meets the ME and MP requirements as well as 12 kg of a 160 g/kg CP concentrate *when the quality of the silage is very good.* The recent claims made for fish meal in high protein concentrates for dairy cows (Table 6.5) do not therefore imply any magic ingredient(s) in fish meal; they are, in fact, exactly according to prediction of ME and MP requirement from first principles of ration formulation.

The relationship between milk yield and MP intake in Fig. 6.3 is based on the assumption that the efficiency of utilisation of MP for milk synthesis (k_p) = 0.62. This is a perfectly acceptable first approximation. However, as explained in Chapter 5, this may simply reflect a response to ME when ME and MP supply are in perfect balance. Figure 5.3 showed, however, that the efficiency of response to increasing MP is only about 20% when ME intake is fixed. A computer-based feed plan based on ME and MP therefore formulates rations that are correctly balanced with respect first to FME and ERDP for the rumen and then to ME and MP for the cow. It calculates requirement for a given milk yield within the constraints of appetite on the assumption that $k_l = k_p = 0.62$.

If, however, one is attempting to increase milk yield, e.g. to meet quota, by increasing the ratio MP:ME then the response to MP alone should be calculated on the assumption that net efficiency is now not 0.62 but 0.20.

Feeding strategies

Feeding programmes for dairy cows in summer and winter are, in practice, influenced by a number of factors that are not directly matters of nutrition, such as the quantity of forage available, the labour and equipment for handling it, facilities for feeding concentrates in and out of parlour, etc. At this stage I only wish to examine how alternative feeding strategies may be used to meet the nutrient requirements of the healthy lactating cow. Implications of feeding strategies for production disease will be considered in the next chapter.

Feeding to yield

The traditional dairy farmer who fed forage for maintenance plus 4 lb of cake per gallon of milk (0.4 kg/litre) was, by definition, feeding to yield, albeit very inefficiently. The modern farmer who feeds according to a computer-based rationing system such as that illustrated in Table 5.8 is also feeding to yield since he is attempting to match the input of ME and MP as precisely as he can to stated allowances, and by so doing ensure that, in theory, he derives the maximum utilisation of home-grown feed.

There may be practical problems associated with feeding strictly to yield. The ration formulation may call for an amount of concentrate in excess of that which a cow may reasonably expect to eat in parlour (about 8 kg/day) and so call for some form of out-of-parlour feeding. The simplest way to do this is to provide a restricted amount of concentrates to cows in mangers out of doors or in their winter accommodation in association with their forage ration. This can be either more commercial cake or a third feed such as sugar beet pulp. The best way to ensure reasonably even distribution of this third feed is to mix it with or sprinkle it on the forage. If it is offered on its own, some cows will inevitably eat much more than others. In particular, low yielding cows in late lactation who are getting very little cake in parlour may gorge themselves out of parlour, which (a) destroys the whole strategy of feeding to yield, and (b) may cause cows to become too fat in late pregnancy. This problem can be reduced (but not eliminated) by offering the third feet at 'lunchtime', i.e. equidistant between the morning and evening milkings when the high yielders should have recovered their appetites.

The second snag with feeding an allegedly fixed ration out of parlour is that the herdsman and his relief milkers still have to check and deliver precise rations of concentrate for each individual cow at each milking. This can be overcome by the provision of a computerised, programmable system for allocating rations in parlour. However, it is easier and more precise to feed fixed rations in parlour (say 2 kg cake at each milking) and provide controlled amounts of extra cake for the high yielders using a computer-controlled out-of-parlour feeder programmed to recognise each individual cow and deliver, in small amounts, cake up to the preordained daily maximum. These feeders have two further advantages:

(1) they prevent cows from gorging too much cake at any one time, in parlour or out;
(2) they provide a daily record of concentrate intake and an early warning of any cow going off food.

They are less popular than they were since, dairy farmers have (rightly) lost enthusiasm for pushing their cows to eat as much cake as possible. In my opinion, however, they are still an excellent aid to controlled feeding in circumstances where a significant number of cows are expected to eat over 6 kg cake/day.

The strategy of feeding to yield is inherently sound but it should not be applied too rigidly because of the inherent uncertainties attached to the estimation of nutritive value and dry matter intake. Moreover, in its most rigid form the strategy is based on the assumption that at all stages of lactation each cow is producing milk exactly to its genetic potential and would not, for example, produce a little more if given a little more food, particularly during early lactation when yield is rising to a peak.

This argument led to the strategy of *lead feeding* which involves giving cows in early lactation an amount of cake corresponding to a milk yield (say) 5 kg greater than current yield, in the expectation of driving milk production up to a higher peak. This policy is also less popular than formerly, partly because of milk quotas but mainly because it did call for very large quantities of concentrates for high-yielding cows in early lactation which (a) did not increase overall gross profit margins of milk sales over purchased concentrate feed, and (b) tended to predispose to disorders of digestion and metabolism. As a general rule, I advise dairy farmers not to exceed the concentrate ration recommended by feeding programmes to meet current yield in early lactation. In other words, I think lead feeding is, in most cases, pointless and potentially dangerous.

After peak lactation milk yield normally declines at a rate of about 1.5% to 2.0% per week. Suppose, for some unspecified and transient reason, the yield of an individual cow fell by 3% per week for two weeks. The farmer feeding strictly to yield would then offer her less cake. Given less nutrients this cow would then be less likely to recover her former lactation pattern. The farmer who feeds less and less as milk yield declines runs the danger of chasing daily yield downhill and preventing cows from revealing their genetic potential for milk yield through the whole lactation.

Brinkmanship

If milk yield, or the rate of decline in yield, appears to be disappointing for some cows or the whole herd, one may assume that nutrient intake is failing to meet requirement and should therefore offer the cows either more cake, if the deficiency appears to involve ME, or cake of a higher crude protein concentration if the deficiency appears to involve MP. (The likelihood of low yields being due to a mineral dificiency in cows getting subtantial amounts of a commercial dairy compound is remote.) However, when cows are performing satisfactorily to their genetic potential it is reasonable to ask the following questions:

(1) 'Could they continue to perform as well given less cake either because the nutritive value of the forage is better than I think or because they are prepared to eat more of it than I predict or both?' (since forage quality and DMI are so closely linked).

(2) 'Can I therefore reduce the cake ration by (say) 2kg/day without incurring a drop in milk yield and so improve my profit margin over concentrate food?'

This approach is called *brinkmanship*. The farmer first feeds to yield, then reduces concentrate intake for individual cows, or group of cows at the same stage of lactation, in increments of 1 kg/day until he reaches the 'brink' and milk yield starts to fall, whereupon he restores cake ration to the previous level. The strategy of brinkmanship, meticulously applied, may save up to half a tonne per cow on concentrate food in a good year compared to a strict feed to yield policy. In a poor year (i.e. when the quality of forage is poor) the careful farmer will probably discover that he cannot reduce concentrate intake without sustaining an unacceptable drop in yield. In either case, the strategy is one that permits him to fine tune rations according to the needs of individual cows or group of cows even though he doesn't know precisely the quality of the food nor the amount being eaten. The practical application of brinkmanship requires quite a lot of planning but it is, in my opinion, the feeding strategy that most elegantly applies the rules of nutrition to optimise the yield of milk from home-grown food.

Flat-rate feeding

In its simplest form, flat-rate feeding means feeding all cows in the herd a fixed ration of concentrates (say 6 kg/day) throughout most of the lactation period. This extreme form of flat-rate feeding is seldom practised, not only because it may underfeed cows in early lactation so preventing them from achieving their genetic potential, but also because some cows may be *overfed* in late lactation and pregnancy, become too fat and experience problems at or shortly after calving (Chapter 7).

A simple variation, namely stepped flat-rate feeding (the '*step-system*') has, however, become very popular and with good cause. Here an estimate is made of the amount of cake required to meet the average milk yield of all cows in (say) the first 16 weeks of lactation, knowing the ME and MP value of the forage. If this is 8 kg then all cows up to 16 weeks of lactation will receive 8 kg of forage per day. Thereafter forage intake may be reduced to, say, 4 kg/day and kept at that level until shortly before the cows are dried off. The number of steps need not be restricted to two; it could realistically be as high as eight.

A simple, stepped flat-rate system has the obvious attraction that it is extremely easy to operate. It can also produce very impressive results. A large number of carefully conducted trials in recent years has shown that flat-rate systems can produce results as good as the most elaborate strategies of feeding-to-yield with brinkmanship in terms of milk yields and margins over concentrates (Owen, 1979; Broster, Johnson and Tayler, 1980; Leaver, 1983). Why should this be?

(1) The flat-rate system tends to avoid excessive intake of concentrates in early lactation and the digestive and metabolic upsets that may ensue.

(2) The shape of the lactation curve can be flattened out somewhat without reducing total yield (i.e. peak yields may be lower but

sustained longer). Moreover, flat-rate feeding does not 'chase yields downhill'.

(3) The best quality forages are so close to concentrate foods in terms of M/D that one can vary concentrate intake over a wide range without significant effect on ME intake (see Fig. 6.2.). (This can, of course, equally well be used as an argument in favour of brinkmanship.)

The stepped flat-rate feeding strategy is simple, healthy and can give impressive financial returns on commercial dairy farms, especially where forage quality is very high. I cannot however accept the argument that it is the 'best' feeding strategy, for a very simple reason. On our farm at Bristol we have individual mature cows with total lactation yields ranging from 3500 litres to over 10,000 litres. A critic might say with some justification that that is a rather sloppy range and we should have been more ruthless in culling our low yielders. However, that's the way it is for us and many farmers in our vicinity. We know that cows giving 10,000 litres a year must be eating, on average, much more forage than that assumed for the purposes of ration formulation. However, we also know that if we cut back on concentrate feeding for the 10,000 litre cow beyond the limits of brinkmanship she will give less milk, but the 3,500 litre cow will not give more milk if we give her more cake.

Feeding the same amount of concentrate at each stage of lactation to cows differing so greatly in their genetic capacity to produce milk must inevitably compromise the capacity of the high yielder to achieve her potential. Overfeeding cake to the low yielder will either make her too fat or discourage her from eating low-cost silage. In either case the economic efficiency of utilisation of the entire ration will be less than if each individual cow were first offered a balance of forage and concentrate appropriate to her yield and then the concentrate ration manipulated by the exercise of brinkmanship to ensure the most efficient utilisation of forage.

Complete diet feeding

The rather meaningless expression 'complete diet feeding' in fact describes a system whereby all the ingredients of the ration (forage, concentrates, mineral and vitamin supplements) are mixed together in a feeder wagon and dispensed into mangers for cows to eat whenever they wish. The various ingredients of the diet are always in the correct proportions for a balanced diet. This tends to encourage high dry matter intakes and stabilise rumen fermentation. The diet is designed to provide all the nutrients available for lactation which, in theory, removes the need to feed cake in parlour. In practice, the cows need something attractive in parlour to persuade them to come in. This could be cake in amounts as small as 1 kg per milking, or even something as simple as a molasses 'lollipop' for the cow to lick while being milked.

Table 6.6 Complete diet feeding of dairy cows. Details of diets used in MAFF trial 1977–78

| | Milk yield | | |
	High	Medium	Low
Composition of DM			
ME (MJ/kg)	11.5	10.9	10.2
crude protein (g/kg)	160	150	140
MAD fibre (g/kg)	220	260	310
D value (in vitro %)	69	64	60
Estimated intake			
dry matter (kg)	18.9	17.4	15.1
ME (MJ)	217	190	153
CP (kg)	3.0	2.6	2.1
Average milk yield (1/d)	25.4	18.0	10.3

Feeding a mixed complete diet almost entirely out of parlour requires a flat-rate strategy or step-system. Table 6.6 gives nutritive value and intake of diets used in complete diet feeding experiments (MAFF, 1978) using a three-step system, M/D for early, mid and late lactation being 11.5, 10.9 and 10.2 MJ/kg DM respectively. This requires the herd to be split into three groups. If cows calve down in blocks they can stay in the same housing and feeding area throughout lactation, the group being demoted to the lower M/D ration when appropriate. Alternatively there can be separate housing and feeding areas for high, medium and low yielders. In this case cows are demoted as individuals as yield declines. This strategy can create behavioural problems as the demoted cows are forced to adapt to the new society and sudden drops in milk yield out of all proportion to the decline in the nutritive value of the diet.

The health of cows on complete diet systems has generally been good. Moreover, there is evidence to suggest that the stable rumen fermentation produced by the balanced mixture of forage and concentrate encourages a high concentration of butterfat. There is however no good evidence to suggest that a switch to a complete diet system will significantly increase total yield or improve efficiency of utilisation of food. The capital required to purchase the feeder/mixer wagon and (probably) to convert existing buildings is high. The dairy farmer who grows only grass and buys in all his concentrates has little incentive to invest in a complete diet system. However for the farmer who grows grass, cereals and perhaps roots, the complete diet system is an excellent strategy for ensuring efficient, healthy and economic milk production from his mixture of home-grown foods.

7 Feeding Problems and Metabolic Diseases

The overworked cow

In Chapter 1 I compared the modern dairy cow yielding 10,000 litres per lactation on a rich mixture of high quality forage and concentrates to a highly tuned racing car designed to run as fast as possible on very high grade fuel; spectacular at best but far more likely to blow up than the old house cow, pottering along at 2000 litres per lactation on grass, hay and minimal amounts of cake.

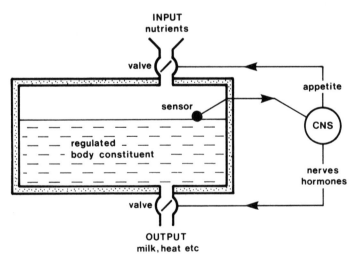

Fig. 7.1 A simple illustration of homeostatic control (for explanation see text).

The physiological control mechanisms of a cow, or any mammal, operate continuously to maintain *homeostasis* or the constancy of all facets of the internal environment of the body such as temperature, body fluid volume and concentrations of electrolytes and substrates such as glucose, amino acids and free fatty acids. Obviously the more rapid the movement of any essential chemical through the body the more rapidly an upset to homeostasis can occur. The principle of homeostasis is illustrated very simply in Fig. 7.1. The quantity or concentration of any regulated body constituent depends on the balance between rates of input and output. In Fig. 7.1 this is represented by a reservoir with its inlet and outlet regulated by valves. The level of the fluid pool within the reservoir, which corresponds to the quantity or concentration of a body constituent such as water, glucose or calcium, is perceived by a sensor and the information transmitted to a controller, the central nervous system, which can maintain the level (or concentration) by regulating both inlet and outlet valves. When the old house cow potters along at 15 litres/day, inlet and outlet rates for most body constituents are low relative to the size of the reservoir and pool size can be maintained within precise limits by relatively minor adjustments to inlet or outlet valves. The 40 litre/day cow has a much higher output rate relative to pool size. In these circumstances:

(1) Fluid level in the reservoir is likely to vary more widely about its set point.
(2) It may be impossible to achieve an input sufficient to maintain fluid level. In this case, the controller (CNS) can only respond by reducing output.
(3) Output control may be insufficient or too slow to prevent catastrophic losses. Postparturient hypocalcaemia (milk fever) is a classic example (see Fig. 2.9(b)). Unless input is massively augmented, e.g. by injection of calcium salts, the animal will die.

Blowey (1985) has used this model of homeostasis to explain the concept of the 'metabolic profile'. This involves measurement of concentrations in blood of important body constituents and metabolites such as glucose, ketone bodies, urea and minerals such as calcium and phosphorus. If the concentrations of a particular metabolite, say glucose, departs significantly from normal this is taken to indicate a disorder of energy metabolism. The relevance of such measurements to the diagnosis of specific metabolic diseases will be discussed in more detail later. At this stage I would point out that abnormal blood concentrations of constituents only imply a disturbance to the equilibrium (a change in water level in the reservoir) and this is not the same thing as production failure. Faced by a chronic input deficiency of (say) copper the controller will slow down output, e.g. growth rate, metabolic rate, to restore equilibrium. Blood copper concentration may be within normal limits but performance will continue to be impaired until copper intake is restored. Metabolic profiles are no more than snapshots of water level in the reservoir; useful, especially if taken during the period of equilibration to a new metabolic demand, but not so useful as would be a precise record of rates of input and output.

A high milk yield, especially in early lactation, imposes a tremendous demand on the capacity of the cow to provide nutrients to the mammary gland. She has the desire to eat more and the farmer the desire to give her more nutritious food. This imposes stresses on the digestive system. To meet the demands of lactation she also mobilises nutrients from her own body reserves, chiefly fat, and this too can place a severe strain on metabolism, especially in the liver.

HAZARDS FOR THE HIGH YIELDER

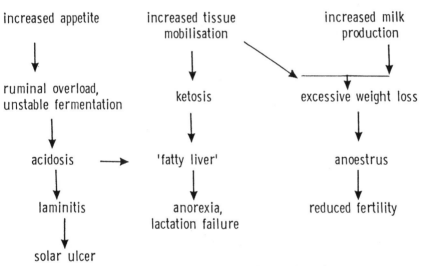

Fig. 7.2 'The overworked cow' – metabolic stresses in early lactation.

Some of the metabolic stresses of early lactation are illustrated in Fig. 7.2. To sustain milk yield the cow increases her intake of food and the farmer increases the ratio of cereal to forage. This is perfectly normal and healthy, up to a point. However, excess intake of rapidly fermented concentrates may lead to indigestion, ruminal acidosis and disruption of the normal ruminal microflora. This will provoke at least acute discomfort, loss of appetite and reduced milk yields. In more severe cases the primary ruminal upset can lead to a general metabolic acidosis. This not only makes the cow feel extremely unwell but can predispose to lameness due to laminitis (acute inflammation of the sensitive laminae of the feet), which causes severe and enduring pain.

Mobilisation of nutrients from body reserves of protein and fat is, once again, perfectly normal and healthy, up to a point. Excessive fat mobilisation can lead to ketosis and fatty degeneration in liver cells. Excessive protein loss can predispose to infertility. Both may increase the risk of other problems post-calving, such as metritis (inflammation of the uterus). I shall deal with these problems in more detail later. The point that I wish to convey here (and by Fig. 7.2) is that most of the disorders of digestion and metabolism seen in early lactation can simply be attributed to overwork. In this chapter I shall consider these and other disorders of the dairy cow that are known to be linked to nutritional stress. I shall not cover problems that cows may encounter through eating harmful substances ranging from fungal toxins on grass or concentrate to pieces of wire, or the lead from old car batteries, nor shall I consider differential diagnosis between these various conditions. Readers seeking answers to such questions of veterinary medicine may consult Blowey (1985) or Blood and Radostits (1989).

Feeding problems

Bloat

Bloat occurs when the normal process of eructation or belching is interrupted and so fails to expel the gases produced by fermentation. The rumen rapidly swells, the cow experiences severe pain and may, in extreme cases, die within a few minutes of showing clinical signs. A distinction can be made between gassy and frothy bloat. In gassy bloat the gas on top of the rumen contents cannot escape either because the cow has lost the ability to belch or, uncommonly, because of some mechanical obstruction to the oesophagus. An obstruction could be caused by the cow choking on a large piece of root crop or by external pressure from an abscess in the throat. The more common cause of gassy bloat, simple failure to belch, is still not properly understood. In the first instance, belching is not completely inhibited but becomes less and less effective as the rumen distends with gas. This condition is typically seen in calves shortly after weaning and in 'barley-beef' animals receiving nearly all their food as concentrates. The rate of gas production is relatively rapid and rumen movements relatively slow due to the lack of long fibre in the diet. In these circumstances a number of individuals may become chronic bloaters whose condition can only be kept at bay by passing a stomach tube to relieve the gas or, in extreme cases, puncturing the rumen wall through the upper left flank with a trochar and cannula. This form of gassy bloat associated with high concentrate feeding is unusual in dairy cows unless, for example, they get into a grain store and gorge themselves. Normally they eat enough long forage to ensure normal rumen movement.

Frothy bloat occurs when the bubbles of gas produced within the mass of rumen contents fail to burst when they reach the top and so combine to form a mass of foam or froth that fills the entire gas space. The surface tension of the liquid holding all these gas bubbles together makes belching impossible. Moreover, the foam cannot be cleared by passing a stomach tube and even a trochar and cannula is sometimes ineffective. Frothy bloat usually occurs in cows at pasture, classically when they have access to large amounts of clover or alfalfa. However, several cows may suddenly develop bloat on a previously blameless pasture for no apparent reason. In this case there is no option but to take them off the pasture and hope that the problem will resolve itself with time.

When a dairy cow bloats at pasture it is reasonable to assume frothing of rumen contents and so administer by drench or direct injection into the rumen an anti-frothing agent. If no suitable veterinary preparation is to hand, 500 ml of linseed oil or liquid paraffin can be used for first aid. With luck, the cow will start to belch at once. In all cases of moderate bloat it is useful to encourage the cow to keep walking, so as to swill the rumen contents around and stimulate belching.

The incredibly rapid progress from onset of clinical signs to death in some cases of bloat has led some scientists to suggest that the cows have been poisoned by toxins released from the rumen. This may sometimes occur but it is probable that most cows die of heart failure as a consequence of (a) intense abdominal pain, and (b) massive pressure on the great blood vessels in the thorax and abdomen obstructing the free movement of blood to and from the heart.

Fog fever

Fog fever describes a condition classically seen in mature cattle shortly after being turned out onto lush autumn pasture. Several cows show severe signs of respiratory distress, standing with their necks stretched and mouths open, straining especially to breathe *out*. I include mention of this disease here because it is largely caused by over-eating. Lush grass, especially when rich in nitrogen, contains high levels of the amino acid tryptophan. In the anaerobic environment of the rumen this is converted to 3-methyl indole which is absorbed into the blood stream and can cause hypersensitive (allergic) damage to the alveoli of the lungs. The condition is more common in suckler cows than dairy cows for the simple reason that the former are more likely to be underfed during the dry months of mid-summer (the 'hungry gap'). When extremely hungry cows are turned onto a new lush autumn pasture they eat avidly and, if sensitised, may poison themselves with 3-methyl indole. Monensin, a common antimicrobial feed additive for beef cattle, has been used with effect in the control of fog fever, although the daily administration of monensin to suckler cows at pasture is easier said than done. Dairy cows can and do get fog fever but the condition is much rarer for the simple reason that they are unlikely to enter autumn pastures in a half-starved condition. If in any doubt, however, access to new pastures should be restricted, e.g. by strip grazing.

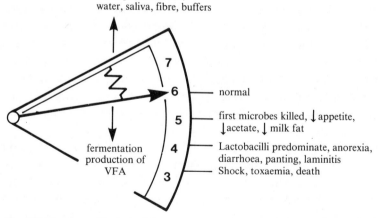

Fig. 7.3 Rumen acidosis and its consequences. The tendency for fermentation to increase acidity (reduce pH) is opposed by buffers and other physiological regulators to keep pH normally at about 6.0.

Acidosis and laminitis

The sequence of events that lead to laminitis is illustrated in Fig. 7.3. In normal circumstances the rumen microorganisms ferment carbohydrates in cell walls (digestible fibre) and cell contents (starches and sugars) to yield volatile fatty acids (VFA) in proportions of about 0.7 acetate, 0.2 propionate, 0.1 butyrate. The tendency for these acids to reduce rumen pH is opposed by the physiological mechanisms for absorbing VFA and for diluting and buffering them with saliva containing salts such as sodium bicarbonate. Normally these keep pace with VFA production rate, so rumen pH remains constant at about pH 6.0. However, if the rate of production of VFA by fermentation exceeds the rate at which they can be absorbed, buffered or washed out of the rumen, then the acidity of the rumen will rise (i.e. pH will fall). The first consequence of this will be destruction, or at least inhibition, of certain cellulolytic species of rumen microorganisms. This reduces the rate of fibre digestion, increases rumen retention time and so inhibits appetite. There may also be at this stage a decrease in acetate:propionate ratio and a fall in milk fat concentration.

Once the rumen pH falls below 5.0 the bacteria and, especially, protozoa associated with stable rumen fermentation start to die in large numbers and lactobacilli, which thrive in more acid conditions, become dominant. They produce lactic acid in an approximately equal mixture of L- and D-isomers, unlike the cow herself (or any mammal) which produces only L-lactic acid in metabolism. These acids are absorbed and blood pH falls. The cow is distressed and pants heavily to lose CO_2 and so increase her capacity to deal with hydrogen ions. D-lactate is a particular problem because it is metabolized much more slowly than L-lactate.

Assuming the cow gets over the acute problems of acidosis she is still left with a ravaged rumen population. The consequences of this are, invariably, depressed appetite and often a foul-smelling yellow-coloured diarrhoea. In very severe cases, usually only associated with gross over-eating (e.g. grain store robbery) cows may die of shock from severe acidosis and absorption of toxins from the rumen.

There are two serious indirect consequences of acidosis, namely ketosis and laminitis. Ketosis (which will be considered in detail later) can follow acidosis if the cow in early lactation loses her appetite and so places a far greater demand on her body fat reserves. Laminitis is an inflammation of the sensitive laminae of the feet. The hooves of the cow, which correspond to our fingernails and toenails, are joined to the sensitive tissues of the foot by complex, folded elastic tissue, the laminae, whose large surface area greatly increases the strength and shock-absorbing properties of the foot (Greenhalgh, McCallum and Weaver, 1981). Following acidosis and possibly the absorption of toxic products from the mass destruction of microorganisms in the rumen, substances are released into the circulation which destroy the lining of the blood capillaries that supply the laminae. This causes progressive destruction of the sensitive tissue that joins the horn to the foot and at the same time stimulates a massive amount of inflammation (i.e. heat and swelling) around the coronet and immediately above the hoof.

This is probably complicated by other mechanical stresses to the foot (Chapter 8). Inflammation and capillary damage behind the walls of the hooves is particularly painful because the swelling has nowhere to go. To imagine how a cow feels when she has acute laminitis, imagine shutting every one of your fingernails in the door and then walking around on your fingertips! It is no exaggeration to say that a cow with laminitis is in excruciating pain. The skin above the hooves is extremely hot to the touch and the cow is extremely reluctant to move. She attempts to take as much weight as possible on her heels. Her back legs may be drawn forward under her body and her front legs may often be crossed. If all four feet are affected, as is often the case, her condition is desperate. Such an animal deserves the relief provided by pain-killers dispensed by a veterinary surgeon. Immediate, if transient, relief can be provided by hosing the feet with cold water.

Since laminitis is such a common sequel to acidosis we may hereafter consider them together. To control acidosis/laminitis in practice it is necessary to ensure that the fermentation rate never exceeds the capacity of the animal to absorb, buffer or dilute the acidic end-products of fermentation. This can be achieved by:

(1) restricting access to concentrates to a maximum of 4 kg per feed (for a Friesian; proportionately less for a Jersey);

(2) giving cows access to long fibre in the form of hay (or possibly good barley straw). This has the disadvantage of reducing overall M/D. However, in circumstances where acidosis is occurring, appetite will already have been inhibited. Such cows will usually eat up to 2 kg/day of hay without effect on their intake of other foods;

(3) partial substitution of foods rich in digestible fibre (e.g. sugar beet pulp) for starchy cereals (e.g. barley);

(4) inclusion of sodium bicarbonate in the diet as a buffer. This can be achieved by inclusion of 10–20 kg/tonne in the concentrate ration, corresponding to an intake of 100–200 g $NaHCO_3$/day. Alternatively, the $NaHCO_3$ may be added to concentrate or forage as fed at an inclusion rate of about 200 g/day.

Although the main cause of acidosis is overfeeding of concentrates, silages have also frequently come under suspicion. The three most common reasons for implicating silages in acidosis/laminitis are:

(1) their pH is very low (below 3.8) and they contain a large amount of lactic acid, especially D-lactic acid;
(2) they have been precision-chopped and so contain insufficient long fibre to stimulate normal rumination, salivation and $NaHCO_3$ flow into the rumen;
(3) they are rich in nitrogen, especially ammonia nitrogen, and this may predispose to the absorption of toxic nitrogenous compounds which may cause laminitis.

I think it unlikely that high quality, low pH grass silage is ever a direct cause of laminitis. In practice, the incidence of laminitis in cows eating a lot of quality silage and relatively little concentrate is low. Precision-chopped silage fed in association with a lot of cereal may constitute a problem due to a shortage of long fibre. In this case, giving cows access to long hay is usually a satisfactory remedy, especially if the nutritive value of the silage is very good (M/D greater than 10.5). If, by eating hay, the cow can avoid acidosis it is likely that she will increase her overall dry matter intake. In this case, hay is a supplementary food, not a substitute.

It is possible that products of abnormal protein degradation in the rumen may cause inflammation and tissue damage in the sensitive laminae of the feet. However, I believe (without proof) that these compounds will not cause laminitis unless the animal is acidotic as a consequence of an unstable fermentation of carbohydrates. The claimed link between laminitis and silages with high ammonia-N concentrations is probably an indirect one. These silages are usually very unpalatable, so forage intake declines with respect to concentrate and therefore the cows may get acidosis/laminitis through eating too much cake.

Maize (corn) silages can be of very low pH (less than 3.8) and still contain large amounts of starch. Such silages can, on occasions, cause acidosis despite their relatively large proportion of long fibre. If this occurs, $NaHCO_3$ at 200 g/day is probably the best remedy.

Abomasal disorders

In the cow subsisting largely on grass or conserved forage the organic material flowing out of the rumen into the abomasum is mostly microbes and the undigested cell walls of plants. The main role of the abomasum is to achieve acid-digestion of microbial protein. However, if carbohydrate fermentation in the rumen is incomplete, the abomasum may become overloaded with partially fermented starch (etc.), abomasal pH remains high so microbes are not killed by acid secretion, and fermentation continues in the abomasum. This generates gas which distends the abomasum and volatile fatty acids which may inhibit its normal ability to contract and propel food into the duodenum. The abomasum is said to become atonic. The condition is most likely to occur early in lactation while the cow is attempting to adapt to more food in general and more starch in particular. It is even more likely when the starch is corn (maize) starch since this is more likely to escape fermentation than starch from barley or wheat. Abomasal indigestion is thus most likely to occur in early lactation. Extreme distension of the abomasum with digestible food can cause abomasal ulcers in young calves. Sudden death can occur (rarely) in dairy cows as a consequence of a haemorrhage from an abomasal ulcer. This may sometimes be due to carbohydrate overload.

In late pregnancy the calf and gravid uterus occupy a large part of the space within the cow's abdomen and this inevitably restricts the volume available to the rumen. After calving and before the lactating cow has built up to peak appetite, there is, in effect, 'room to spare' within the abdomen. If the abomasum becomes atonic or distended by gas while the rumen is relatvely small compared to total abdominal space, it may become displaced from its normal position on the right flank and squeeze under the rumen so that part of the abomasum becomes trapped between the rumen and the *left* abdominal wall. Left-sided displacement of the abomasum is therefore seen most commonly in large-bodied, mature cows almost exclusively within the first six weeks of lactation. It is nearly always associated with the feeding of diets rich in cereals and lacking in long fibre and is practically unknown in cows at pasture.

The clinical signs of displaced abomasum are rather variable depending on the extent and permanence of the displacement. Food intake is reduced or stops altogether. There is a sharp fall in milk yield. The affected cow rapidly becomes ketotic (see below). The diagnosis and treatment of displaced abomasum are matters for the veterinary surgeon. It can often be replaced by external manipulation but the condition is likely to recur since the original displacement usually tears some of the original attachments. Such cases require surgery and this, too, is outside the scope of this book. The *prevention* of displaced abomasum is, however, once again, largely a matter of proper feeding in the first weeks of lactation.

Metabolic disorders

Ketosis and fatty liver

Ketosis, sometimes called acetonaemia, is a very common disease of potentially high-yielding cows due to a failure to balance the metabolic demands of early lactation. To understand the causes of ketosis it is necessary to invoke a little chemistry. Figure 7.4 illustrates the final common pathways of the main substrates used for energy metabolism in the ruminant. The metabolism of organic compounds to make energy available as ATP first involves the splitting up of the large molecules into fragments containing two or three carbon (C) atoms. Fatty acids from the body fat depots break down into 2C fragments. Fermentation in the rumen yields mostly 2C fragments as acetate (and some butyrate). Propionate is a 3C fragment. Glucose, which is

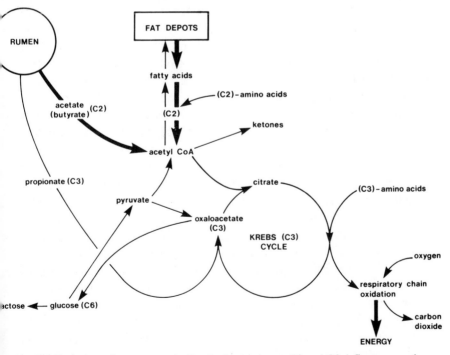

Fig. 7.4 Pathways of energy metabolism in the ruminant. C2 and C3 define two- and three-carbon compounds.

essential for life, is made up of two 3C fragments. The nature of rumen fermentation and digestion is that little, if any, glucose is absorbed from the gut so that the ruminant has to synthesise its own glucose from C3 fragments arising from propionate or certain amino acids such as glutamate, which are therefore called *glucogenic*.

The final common pathway for oxidation involves the combination of 2C and 3C fragments as acetyl coenzyme A and oxaloacetate respectively, to form citrate. This proceeds through a series of intermediate reactions to make available chemicals which can react with oxygen to make energy available for the body as ATP (Chapter 1). The process regenerates oxaloacetate which can then react with more acetyl CoA to continue what is usually called the Krebs cycle, after its discoverer, Sir Hans Krebs. His own title, the tricarboxylic acid (i.e. C3), is more informative.

At the onset of lactation the specific demand for glucose is greatly increased to meet the capacity of the mammary gland to synthesise milk lactose. At the same time the overall metabolic demands for synthesis of milk fat and protein and the work of maintenance and milk synthesis stimulate the cow to mobilise energy from her reserves of body fat. This comes in C2 fragments (see Fig. 7.4). In early lactation, therefore, C3 fragments are being driven towards lactose, via glucose, and fatty acids are being broken down at an increased rate to acetyl CoA. In order to enter the Krebs cycle each molecule of acetyl CoA must be matched by a molecule of oxaloacetate. However, oxaloacetate is regenerated through the reactions of the Krebs cycle. When the rate of production of acetyl CoA from acetate, butyrate, fatty acids from body reserves and 2C (ketogenic) amino acids is high, and when a substantial fraction of oxaloacetate is required for glucose synthesis, then some acetyl CoA cannot enter the Krebs cycle and is diverted into the production of ketones (acetone and beta-hydroxybutyrate). Small quantities of ketone bodies can be used in energy metabolism or excreted. High concentrations are poisonous.

The cow with ketosis in early lactation becomes progressively dull and lethargic. She loses her appetite, usually refusing concentrates before she refuses forage. This is a paradox because starchy concentrates generate more C3 propionate than forage and this ought to improve matters by generating more oxaloacetate (see Fig. 7.4). Milk yield declines progressively. In severe cases the cow loses weight rapidly and becomes extremely dehydrated. In other cases, cows may show acute nervous signs such as blindness and delerium; bellowing, pressing their head against the wall, walking in circles. Many mild cases of lactation ketosis in dairy cows are self-limiting as milk production declines to the level at which the supply of acetyl CoA and oxaloacetate get back in balance. The condition is similar in metabolic terms to pregnancy ketosis or 'twin-lamb disease' in sheep, which results from the high demand for glucose from twin lamb foetuses at a time when food intake is depressed by the space occupied by the uterus. In pregnant sheep the condition is frequently fatal since the ewe cannot reduce glucose demand except by aborting her lambs. Ketosis seldom kills dairy cows but it ruins their productivity.

Almost every cow that gets ketosis will suffer to some degree from the condition known as 'fatty liver'. Ketone bodies are manufactured in the liver although oxidised elsewhere. The liver also manufactures lipoproteins which are used to transport fats throughout the body. If the liver becomes swamped with ketones and fatty acids from body reserves some liver cells are damaged or killed and others lose their functional capacity as they fill up with fat. This liver malfunction is undoubtedly the main cause of depression and inappetence in the ketotic cow.

PREVENTION AND TREATMENT

To prevent ketosis and the fatty liver syndrome it is necessary to avoid an accumulation of C2 fragments in early lactation. One way to achieve this is to ensure that cows are not too fat at calving. The amount of fat and muscle cover on a cow is usually assessed according to a 'condition score' from 1–5, where 1 is emaciated and 5 is obese (for pictures and detailed description see Fig. 12.3). There are some who advocate that to avoid ketosis cows should naturally be slightly thin at calving (average condition score 3 or less). In my opinion this is a little too thin for optimal performance. In common with Blowey (1985) I think a cow at calving should be fit but not fat (condition score 3–3$\frac{1}{2}$). Cows that are too fat in late lactation tend to have low appetites especially for forage because the large amount of intra-abdominal fat allows even less space for the rumen. Inappetence in early lactation obviously predisposes to ketosis because food, via propionate, is the main source of C3 units.

The role of protein is equivocal. Glucogenic amino acids are alternatives to propionate as sources of C3 fragments. Indeed, substantial amounts of dietary protein and body protein reserves are used by cows to make glucose rather than milk proteins. However, there are some convincing reports that high protein concentrate rations may exacerbate the problem of ketosis in practice. I am prepared to believe this but don't understand it. Maybe the high protein rations simply don't contain enough starch.

One thing that can be said with certainty is that the incidence of ketosis is much greater in cows permanently tethered in barns in those given reasonable exercise (Ekesbo, 1968). Healthy exercise stimulates appetite but prevents obesity. It also stimulates the hormones of the adrenal cortex – the glucocorticoids – to mobilise body reserves, including the glucogenic amino acids from body proteins to meet the modest metabolic demands of exercise, and so 'trains' them to meet the much more severe metabolic demands imposed by the onset of lactation. This should also explain why adrenal steroid hormones are given by injection as part of the treatment for ketosis.

The most obvious therapy is to make more C3 units available to the cow. Oral administration of glucose is of little use since it will be rapidly fermented to VFA. An intravenous infusion of glucose can produce a rapid improvement if the cow's liver is not too damaged by fatty infiltration, but many cows relapse. An interesting new form of therapy involves the oral administration of monensin (25 mg/day), an antibiotic used in great quantities as a food additive for beef cattle for its ability (amongs other things) to increase the ratio of propionate to acetate yield from rumen fermentation. Increasing propionate improves the ratio of C3:C2 fragments absorbed by the ketotic cow and can apparently produce a rapid return to good health. Monensin depresses butterfat and reduces appetite in normal cows so has no role as a food additive for milk production.

ASSOCIATED CONDITIONS

There is a proven association between ketosis/fatty liver and other disease conditions in early lactation including retained placenta and metritis (inflammation of the uterus), infertility, milk fever and even mastitis and cystic ovaries. This has led to the claim that the ketosis/fatty liver complex may be an important precipitating factor in these conditions. It is difficult to construct a logical interpretation to support the claim that ketosis may lead to (e.g.) cystic ovaries. On the other hand it is much easier to understand how primary stresses in early lactation, such as metritis and mastitis, would cause a cow to become ketotic when simultaneously faced with the stress of lactation. One has to be cautious in assigning what is cause and what is effect. Nevertheless, conditions such as retained placenta and cystic ovaries do appear to be more prevalent in cows that are too fat at the time of calving. Thus, for whatever reasons, the cow that is fit but not fat at calving, and well adapted to the diet that she will get in early lactation, is less likely to suffer not only from ketosis but from a lot of other problems too.

Parturient hypocalcaemia ('Milk fever')

The condition known to every dairy farmer as 'milk fever' appears in veterinary textbooks as parturient hypocalcaemia or parturient paresis. The scientific titles are ponderous but correct, the popular title 'milk fever' being memorable but wrong. The condition is caused by a severe drop in the concentration of calcium in the blood and extracellular fluid (ECF). Calcium is essential for normal function of muscles in the skeleton, heart and gut. Its concentration in the ECF is regulated by parathyroid hormone and vitamin D_3 (cholecalciferol) which control (a) net uptake of calcium from the gut, and (b) mobilisation of calcium from the skeleton. At the time of parturition the dairy cow suddenly begins to synthesise a large quantity of milk in the mammary gland and demands calcium from the ECF at a rate that could, if not replenished, exhaust her reserves of freely available calcium within 10 hours (Fig. 2.9(b)). She is, moreover, likely to eat very little on the day that she calves. It is inevitable therefore that blood calcium concentration will fall to some degree during parturition and the onset of lactation. If the fall becomes too severe, the muscles of the cow become progressively weaker (paresis) and she is unable to stand. In some cases this weakness is preceded by a period of excitability with muscle tremors. When she 'goes down' she initially, at least, adopts a reasonably normal posture, lying on her brisket but often with a characteristic S-shaped bend in her neck. Later her head may fall against her chest. Her anal region becomes swollen with faeces that her gut is too weak to expel. Unless treated her condition rapidly deteriorates; she shows signs of shock, her ears, muzzle and limbs become cold to the touch. Finally she falls onto her side, becomes progressively bloated and dies probably of respiratory or cardiac failure.

An incidence of milk fever of between 5% and 8% in dairy herds is considered 'normal' – which is a fairly alarming definition of normality. Parturient hypocalcaemia is almost exclusively a problem of high-yielding dairy cows, being rarely seen in suckler beef cows at pasture. Beef cows secrete less milk. Moreover, their calves are unlikely to drink more than about 6 litres/day for the first few days of life. Thus in truly normal circumstances the build-up of demand on mechanisms for increasing calcium absorption from the gut and mobilisation from the skeleton are much more gradual.

Some cows are more prone to milk fever than others with similar milk yields, which suggests a genetic predisposition. The condition is particularly common in Jersey cows. The incidence also increases with age since older cows are less able to mobilise calcium from the skeleton. It can also be produced experimentally by feeding pregnant cows diets with a high calcium:phosphorus ratio (Ca:P). Since parathyroid hormone increases uptake of calcium and execretion of phosphorus, it is assumed that such diets cause the parathyroid gland to become sluggish and so fail to mobilise sufficient calcium when needed at the onset of lactation. In practice, the incidence of milk fever tends to be high when cows graze large quantities of lush spring or autumn grass. Grass has a high Ca:P ratio and may inhibit parathyroid hormone secretion. This is probably an incomplete explanation since the problem does not appear to be particularly severe for cows eating lucerne which has an extremely high Ca:P ratio. Grass which has a high concentration of nitrogen, a low concentration of magnesium, or is simply very wet, may impair uptake of calcium from the gut. There is also a suggestion that the incidence is higher in cows that are too fat at calving or have been *excessively* 'steamed up' with large amounts of concentrate food during the last weeks of pregnancy. Using the logic of Fig. 2.9(b) we might cautiously assume that fat cows are more likely to lose their appetite at calving, especially if they become ketotic; excessively 'steamed-up' cows may be too fat or produce extremely large quantities of milk in early lactation. In either case this would tend to overstress calcium homeostasis.

PREVENTION AND TREATMENT

A condition that may affect 10% of the dairy herd and can kill a cow within a few hours of the onset of clinical signs would be catastrophic were it not for the fact that diagnosis and treatment are usually simple. Most cows respond to the intravenous injection of 6–12 g calcium usually as 100–200 g calcium borogluconate in a 25% solution (i.e. 400–800 ml of solution), which is best delivered by slow intravenous injection since this produces a rapid response and no tissue damage. Subcutaneous injections may be administered by the farmer:

(1) as a preventive measure in a cow that has had milk fever at previous calvings;
(2) to prevent relapses following a first intravenous injection. The cow that fails to rise following treatment with calcium borogluconate is not suffering from simple parturient hypocalcaemia and requires the skilled attention of a veterinary surgeon. She may respond to a mixture of calcium, magnesium and phosphorus or may be suffering from other problems post-calving. The 'downer cow' is discussed later.

It is not easy to devise an effective strategy for preventing milk fever, which is why the incidence remains high in a number of well-managed dairy herds. Experimentally, cows in late pregnancy can be fed diets relatively low in calcium and with a Ca:P ratio of 1:1, say 30 g/day of each, in order to stimulate parathyroid activity. In practice, cows at grass are likely to be consuming about 100 g calcium per day and no dairy farmer wants to deny his cows access to good grass. While it may be impractical to reduce calcium intake, there is a case for increasing both phosphorus and magnesium intake in late pregnancy. These minerals may be added to the drinking water or included in a mineral block. The latter solution is not ideal because cows lick mineral blocks in an erratic fashion and cannot be relied upon to consume minerals according to requirement. In my opinion the best solution for problem herds is to incorporate appropriate quantities of phosphorus and

Table 7.1 Recommended constituents in mineral supplements for cattle at pasture (g/kg)

	High magnesium	High phosphorus
Salt (NaCl)	500	240
Magnesium	180	15
Calcium	–	150
Phosphorus	–	120
Potassium	0.18	–
Iron	1.30	1.30
Cobalt	0.13	0.13
Copper	–	1.25
Selenium	–	0.008
Iodine	0.15	0.15
Zinc	1.16	1.40
Vitamin D_3	–	55000 iu

magnesium into a concentrate food used to 'steam-up' the cows prior to the onset of lactation. I have stated earlier that I believe it is in the best interests of both cow and farmer to adapt her to concentrate feeding prior to the onset of lactation provided that she does not become too fat or suffer from premature let-down of milk prior to calving. I would recommend not more than 4 kg/day of a compound food containing 140 g/kg crude protein, 10 g/kg phosphorus, 2 g/kg magnesium and a generous amount of vitamin D_3. This may mean separating the dry, pregnant cows from the milking herd, but so much the better (see Chapter 8). Alternatively a high phosphorus, high vitamin D_3 mineral supplement (Table 7.1) may be sprinkled onto the food or offered free choice.

It is also possible to reduce the incidence of milk fever by the injection of heroic doses of vitamin D (about 10 million international units) about 8–10 days before calving. Cows who have previously gone down with milk fever receive a subcutaneous injection of 400 ml of a 25% solution of calcium borogluconate immediately after they have calved. It also makes sense to draw off no more colostrum than strictly necessary for the calf in the first day and not milk the cow right out for at least two days.

The downer cow

The 'downer' cow is one who goes off her feet, usually at or shortly after calving, and cannot get up. I include the condition here because it is often (but by no means always) precipitated by milk fever. Other causes include:

(1) Generalised weakness following a difficult calving, possibly with excessive blood loss.
(2) Nerve damage during calving; especially damage to the obdurater nerve. Affected cows lose control of the internal muscles of the thigh and do 'the splits' when forced to rise.
(3) Fractures to the pelvis or hind limbs, possibly incurred because of muscle weakness brought on by hypocalcaemia.
(4) Severe pain and illness from acute diseases such as mastitis.

If the downer cow lies in one position for too long, matters deteriorate. Anybody whose arm or leg has 'gone to sleep' after an hour or so of sitting or lying in the same position will appreciate the problem for the nerves and muscles in the limbs trapped between 600 kg of recumbent cow and the hard ground. Unless she is moved, severe destruction of nerve and muscle tissue will begin within 3–4 hours. It is vital therefore to get the downer cow onto a soft bed. This is easier said than done but see Blowey (1985). Her lying position should be adjusted to place weight alternately on the left and right sides, ideally about six times daily. If she is bright but unable to use her back legs, she

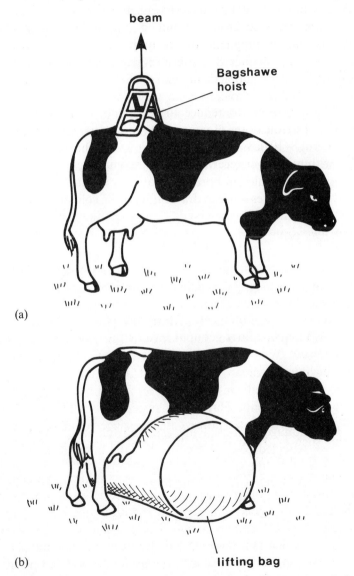

(a)

(b)

Fig. 7.5 Methods for getting a 'downer cow' to her feet. (a) The Bagshawe hoist. (b) The lifting bag.

should be supported in a standing position for short periods either by use of the Bagshawe hoist which lifts her hips or by inflating a lifting bag underneath her (Fig. 7.5). The latter approach looks kinder than the former but is not necessarily more effective.

I have a great deal of sympathy with downer cows and clearly most of them are genuinely unable to use their back legs but I cannot escape the conclusion that some of them (unlike horses) don't *try*. Some cows will remain slumped over a lifting bag but do attempt to help themselves when supported by a Bagshawe hoist. I believe the veterinary surgeon needs to have both lifting aids to hand and use them according to his diagnosis of the individual case.

Recovery can take anything from a few hours to over 10 days. Some cows never get up. It is difficult for the vet to forecast which cows will recover and so advise the farmer whether to persevere with nursing or make arrangements for humane emergency slaughter (see Chapter 8). What he can guarantee is that if she is not nursed properly she will never get up.

Hypomagnesaemic tetany ('grass staggers')

Magnesium has a controlling, inhibitory effect on nerve and muscle activity. In acute magnesium deficiency this inhibition is removed and the cow simply goes mad. The condition is usually sudden in onset and, unless treated, proceeds rapidly to death. An affected cow first appears restless and hyperexcitable. She may bellow or gallop round the field for no reason. Wild-eyed and frothing at the mouth, she becomes increasingly uncoordinated, staggers and falls to the ground where she may lie rigid in total muscle spasm or thrash out with her legs in convulsive movements. Having exhausted herself she may then lie quietly but the whole mad process may be triggered off again by a mild stimulus such as the arrival of the farmer or vet. Death may occur within an hour of the onset of clinical signs and cows with hypomagnesaemic tetany are often found dead. The condition can usually be distinguished from other causes of sudden death (like lightning strike) by the amount of grass and earth stirred up by the cow in her final convulsions. Although hypomagnesaemia usually takes this peracute form, cows which are marginally deficient in magnesium can show intermittent signs of nervousness, tremor and irrational behaviour, especially when provoked. Appetite and milk yield are also reduced. If these clinical signs are accompanied by low concentrations of magnesium in blood (below 18 mg/litre) the herd should be treated before things get worse.

The flow of magnesium (Mg) through the body of a cow yielding 30 l milk/day is shown in Fig. 7.6. Intake is 30 g/day. The apparent

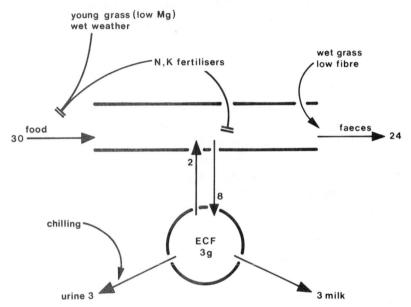

Fig. 7.6 Exchanges of magnesium in a cow yielding 30 l milk/day. The extra-cellular fluid (ECF) contains 3 g. Exchanges are in g/day. K = potassium, N = nitrogen, Mg = magnesium.

absorption is only 20%, so 24 g/day appears in the faeces. The cow appears to have no physiological mechanism for regulating net uptake of Mg from the gut or mobilisation of Mg from the skeleton. Moreover, as with calcium (see Fig. 2.9), the amount of freely available Mg in the exatracellular fluid (ECF) is very small (3 g) relative to the daily movement of Mg across the gut wall and into milk and urine. Thus the cow can run out of Mg fast.

The absence of a physiological mechanism for controlling the supply of Mg to the ECF suggests that Mg deficiency did (and does) not present a great problem of survival to wild ruminants, which further suggests that hypomagnesaemic tetany is an acquired problem of domestication and intensification. Mixed, unimproved pastures nearly always contain sufficient Mg for maintenance and low levels of production. The Mg concentration in grass is lowest in the spring when the grass is young and leafy, especially if heavily fertilised with nitrogen and potash (K), both of which inhibit Mg uptake by the plant. Grasses rich in N and K form insoluble complexes with Mg in the rumen thus inhibiting Mg uptake from the gut (Fig. 7.6). Moreover, young, lush, wet grass, lacking long fibre, also leads to large amounts of wet faeces and so increases Mg loss. (An increase in faecal output associated with a lack of long fibre appears paradoxical. It occurs because the rumen retention time of grass, and other foods, is decreased.)

In the cool, temperate, grassy dairy country of north-west Europe, hypomagnesaemic tetany is typically seen in dairy cows shortly after turn out onto high-quality spring grass. Beef cattle and other less productive animals kept out on poorer land throughout the year are more likely to experience the condition in autumn and winter. In each case individual cases of peracute hypomagnesaemic tetany are often precipitated by a period of bad weather, especially rain or snow. Although cows are extremely tolerant to dry cold, wet and windy conditions can be particularly stressful because they break up the insulation of the coat. The cow that shivers requires more magnesium. The main problem, however, is that in conditions of wind and driving rain or snow cows typically stop grazing and stand motionless with their backs to the wind and their heads bowed. This is designed to reduce water penetration of the coat in the same way that a man caught out in the rain in mackintosh or oilskins stands still to prevent the rain running down the back of his neck. Whereas this may be a less uncomfortable option for the cow than continuing to graze through the rain, she cannot afford to interrupt the supply of dietary Mg to the ECF for more than a few hours. The dairy cow turned out to grass in the spring may become marginally deficient in Mg because the grass is low in Mg but high in N and K, which reduce Mg availability, and she is losing excessive amounts of Mg in wet faeces. If, in this high-risk condition, she then elects to go without food for 10 hours during a stormy night this may be enough to tip the balance.

PREVENTION AND TREATMENT

Acute hypomagnesaemic tetany may, if caught in time, be successfully treated by infusing magnesium salts. Intravenous infusion is, however, very risky and should not normally be attempted by the farmer since a sudden increase in Mg concentration can stop the heart. However, the progress of the disease is so rapid that by the time the vet arrives it may well be too late! Farmers should have an emergency supply of a

25% solution of magnesium sulphate and administer 400 ml *subcutane-ously* at the first onset of clinical signs. This may be followed by oral administration (drenching) of 60–90 g of calcined magnesite in solution.

In this disease almost above all others, prevention is the most cost-effective solution even if it only saves one good cow a year. Several approaches have been recommended; only some of them are realistic:

(1) Spreading calcined magnesite or similar compounds on pastures is expensive, inefficient, and tractors can do terrible damage to a wet spring pasture.
(2) High magnesium mineral licks (Table 7.1) or magnesium/molasses 'lollipops' are useful for beef cows, but, as with all minerals offered free choice, intake is erratic and not proportional to requirement.
(3) Dosing cows with magnesium 'bullets' which, nominally, lodge in the reticulum and release controlled amounts of Mg, is very effective for most cows but some individuals lose their bullets and one dead cow is too many.
(4) If the cows normally drink from a trough, magnesium can be added in controlled fashion to the drinking water. There are commercial Mg dispensers which normally work well but, like all gadgets, need to be checked daily.
(5) The lactating cow is still likely to be getting some cake while at grass. All compound food manufacturers produce a high Mg, high fibre spring grazing nut and this is an excellent way of ensuring controlled, generous Mg intake during high-risk periods.

In general I believe that hypomagnesaemic tetany is severe enough to justify a 'belt and braces' approach: e.g. Mg supplemented drinking water for all cows and heifers, Mg rich cake for those cows in milk and possibly Mg bullets for heifers and dry cows getting no cake.

Other mineral deficiencies

A great deal has been written about mineral deficiencies in grazing animals, especially phosphorus and the trace elements, copper, cobalt and selenium. There is no doubt that deficiencies of all these elements can cause disease and serious loss of production in cattle who are expected to obtain most or all of their nutrients from grass, since grass is seldom, if ever, perfectly balanced with respect to minerals.

Moreover, the tendency in recent years to increase yields of grass per hectare by increased application of inorganic fertilisers (N, P and K) has inevitably tended to reduce the concentration of trace elements such as copper, expressed per kg grass. In dairy herds, the animals most likely to suffer from deficiencies of these minerals are the in-calf heifers since these are the animals that will have eaten most of their food in the form of fresh or conserved grass. Compound cakes for lactating cows are generously supplemented with essential minerals so adult cows are unlikely to experience problems associated with deficiencies of phosphorus or the trace elements unless fed large quantities of 'straights' without mineral supplements. For this reason I shall deal with these problems only briefly.

Phosphorus

The main mineral constituent of bone is calcium phosphate and about 75% of body phosphorus is contained in the skeleton. Most of the remainder is intimately linked to the processes of energy metabolism, principally as adenosine triphosphate (ATP), the energy currency of the body. The first signs of phosphorus deficiency are therefore a non-specific slowing down of metabolism, reduced growth rate and reduced appetite. Many authors have linked infertility in beef cattle on open range to phosphorus deficiency. There may be a direct link but it is more likely that infertility is a secondary consequence of loss of appetite and body condition in phosphorus-deficient animals. I know of no good evidence for a primary link between phosphorus deficiency and infertility in high-yielding cows. Osteoporosis (bone demineralisation and weakness) can be caused by phosphorus deficiency but once again this is usually only seen in unproductive cattle in poor condition getting little food of any sort. It is hardly likely to affect the cow fed for high yields.

Grasses, on the whole, tend to be marginal or slightly deficient in phosphorus (3 g/kg DM or less) and the concentration falls as the plant matures. Deficiencies are most likely during the winter in the higher latitudes and during the dry season in the tropics. Whole cereals are usually adequate, and vegetable by-products such as soyabean meal and brans rich in phosphorus. Most compound foods or feed blocks contain inorganic phosphates.

It is difficult to make a specific diagnosis of phosphorus deficiency, although blood concentrations below 40 mg/litre would be good grounds for suspicion. It has been claimed that cows with phosphorus deficiency develop 'depraved' appetites, eating soil, bones, etc. Some interpret this as proof that cows have a specific ability to seek out and recognise dietary sources of phosphorus, although this has not been confirmed experimentally. Still others claim that primary phosphorus deficiency never occurs in practice (in the UK at least). I don't know what to believe. However, I have seen cases of ill-thrift and infertility in heifers who have been at pasture almost all the year round in the gentle climate of south-west England, and who have responded to high phosphorus mineral supplements (Table 7.1). Moreover, the animals consumed these supplements avidly – though whether that was because they specifically recognised the phosphorus, or the supplement was especially palatable I cannot say.

Copper

Pastures are considered to be copper deficient if they contain less than 10 mg/copper per kg DM. The metabolic roles of copper in cattle are listed in Table 7.2. As with most mineral deficiencies the initial symptoms are non-specific – ill-thrift and inappetence. Diagnostic signs like depigmentation of hair, especially around the eyes (grey 'spectacles') are usually only seen after animals have lost condition. This also applies to swelling and (occasionally) fractures of the distal metacarpal and metatarsal bones just above the fetlocks. Blood copper concentrations are not particularly helpful in reaching a diagnosis. However, copper is stored in the liver and, in rare cases, a liver biopsy may be justified.

A secondary copper deficiency can be induced if cattle graze pastures rich in molybdenum (over 10 mg/kg) and sulphur. In the anaerobic conditions of the rumen these combine to form insoluble copper thiomolybdate. This problem of these teart pastures has been recognised for over a thousand years in my county of Somerset.

Table 7.2   Metabolic roles of copper and signs of deficiency in cattle

1. synthesis of red blood cells	anaemia
2. synthesis of melanin	depigmentation, 'spectacles'
3. bone growth	bone swelling and fractures, especially above fetlocks
4. tissue respiration	ill-thrift, inappetence

Affected animals display the ill-thrift and specific symptoms of copper deficiency but this is complicated by a chronic profuse grey diarrhoea ('teart scour'). For reasons as yet unknown to me, conserved silage and hay from the teart pastures cause less problem than the fresh grass.

Copper supplements for cattle need to be administered with care. If copper-rich mineral licks are offered free choice, individual consumption is extremely erratic and bears no relation to requirement. Such licks must be kept away from sheep who are particularly susceptible to copper poisoning, which can mean sudden death. Acute copper poisoning is not a problem in cattle but dietary concentrations in excess of 40 mg/kg (uninhibited by high concentrations of molybdenum or sulphur) can inhibit the rumen microbes and so reduce appetite. There are several commercial preparations for oral administration which are designed to lodge in the reticulum or abomasum and provide a slow release of copper (plus, probably, cobalt and selenium). These are very good so long as they stay in the gut, which happens in most, but not all, cases. Since copper is stored in the liver, the most reliable way to ensure an adequate controlled supply is by injection at appropriate intervals (perhaps 2–3 times per year).

The profuse grey scour that occurs in cases of teart may be due to a toxic effect of molybdenum within the gut. In this case it is therefore logical to provide a source of copper in the rumen (e.g. a slow release bolus) to bind the molybdenum as thiomolybdate and so prevent the scour, *plus* a secondary supply of copper downstream of the rumen (e.g. by injection) to prevent the symptoms of primary copper deficiency.

Cobalt

Cobalt is the essential mineral constituent of cyanocobalamin or vitamin B_{12}. The rumen microbes synthesise vitamin B_{12} which is then absorbed by the cow. A vitamin B_{12} deficiency causes anaemia in all species. Ruminants have an additional need for B_{12} to support glucose synthesis from propionate (see Fig. 7.4). Cobalt (and thus B_{12}) deficiency jams a large spoke in the wheel of intermediary metabolism. Affected animals suffer a severe loss of appetite and 'pine', or rapidly lose condition. The cobalt-deficient pastures of the world are well recognised. Cobalt supplementation must be by mouth since the element is required not by the cow but by the rumen microbes. Long-acting bullets or glass boluses are the carrier of choice for cattle at pasture with no supplements of cake.

Selenium

Selenium is incorporated in the enzyme *glutathione peroxidase* whose metabolic role, in association with vitamin E, is to prevent the accumulation of harmful organic peroxides as a consequence of oxidation reactions in, especially, muscle membranes. The dietary requirement for selenium is extremely small (about 0.1 mg/kg DM) but it appears that an increasing proportion of pastures are, or are becoming, deficient. However, the element is also extremely toxic in excess and the safe margin between deficiency (less than 0.1 mg/kg DM) and toxicity (over 5 mg/kg DM) is extremely narrow. It is an element that should, therefore, be treated with great caution. A deficiency of selenium and vitamin E, known as 'white muscle disease', can occur in young calves at grass. This presents as ill-thrift, stiffness and, especially, extreme weakness of the shoulder muscles so that the shoulder blades stick out above the line of the spine.

There is some evidence that dairy cows with low blood selenium, or glutathione peroxidase, levels are more likely to suffer from retained placenta after calving. While this is unproven, it may be worth giving controlled amounts of selenium and vitamin E by injection to cows prior to calving in herds where the incidence of retained placenta is high and no other cause can be found.

Cattle also have absolute dietary requirements for sodium, chlorine, iron, iodine, zinc and manganese (Agricultural Research Council, 1980). While these are essential to the cow they are not particularly important to the reader since it is most unlikely that any dairy cow is likely to go short of these minerals under normal husbandry conditions.

Vitamin deficiencies

Vitamins are organic compounds needed in small quantities for metabolism which cannot be synthesised by the body and so must be provided in the diet. This general definition needs some qualification when applied to ruminants. The water-soluble compounds such as thiamine, riboflavine, biotin, cyanocobalamin, etc., which are recognised as members of the vitamin B complex for man and simple-stomached animals, are not true vitamins for ruminants since they are synthesised in adequate amounts by the rumen microbes. Vitamin C (ascorbic acid) is not required by ruminants either. There are some (usually the manufacturers) who recommend mixed B vitamin supplements for dairy cattle. The only circumstance in which they are likely to have any benefit at all is when a cow has not eaten for several days and the population of rumen microbes has become very small.

Vitamin B_{12}, cyanocobalamin, is a special case. Once again there is no justification for using it as a regular dietary supplement but injections of B_{12} do seem to stimulate appetite in cows with ketosis or non-specific inappetence. This can be explained, in part, by its direct effect on gluconeogenesis but I think it has other, unknown, tonic effects. In any event, anything that encourages a ketotic cow to start eating helps to put her on the road to recovery.

The fat-soluble vitamins A, D and E are all essential for cattle. However, the cow usually does not consume the actual vitamins themselves, since they are more commonly found in animal tissues, but a chemical precursor or provitamin which is converted to the active vitamin after absorption into the tissues of the body.

Vitamin A

Vitamin A, retinol, exists only in animal tissues. Cattle obtain their vitamin A largely from the provitamin beta-carotene in grass and other green foods. Cereals (except yellow maize), hay and straw are deficient in beta-carotene. Severe deficiency of vitamin A can cause blindness. Marginally deficient animals run an increased risk of infection of the mucous membranes of the gut and respiratory tract because vitamin A is essential for their maintenance and repair.

Vitamin A is stored in the liver and a cow that has been at grass for six months may well store enough to meet her own needs throughout the winter. Moreover, compound rations for dairy cows are normally supplemented with vitamin A to be on the safe side but the vitamin A status of cows does tend to fall during winter. This may do her no harm but can cause problems for her calf. The colostrum from cows calving in late winter may contain insufficient vitamin A to protect calves from enteritis and pneumonia (Webster, 1985).

Other than young calves, the only other classes of cattle likely to suffer a vitamin A deficiency are growing stock overwintered on some combination of hay, straw, cereals or roots. Silage is probably not as good as grass as a source of beta-carotene, but good enough.

Vitamin D

Vitamin D is a generic name for a range of chemical compounds called *calciferols* which, as already described, regulate the absorption of calcium from the gut and the concentration of calcium in the extracellular fluid (see Fig. 2.9). The main source of vitamin D for cattle is the provitamin *ergosterol* in green plants which is converted by ultraviolet light to ergocalciferol (D_2). Sun-cured hay can be rich in vitamin D_2. However, cattle, like ourselves, synthesise D vitamins from provitamins when outdoors during daylight.

Most of the time the vitamin D requirement of cattle is small because their diet is rich in calcium. Young stock, overwintered in sunless hovels on poor quality silage or cereal and straw may benefit from supplementation (probably by injection) with vitamins D and A. The onset of lactation dramatically alters calcium metabolism (see Fig. 2.9). Injection of massive amounts of vitamin D_3 prior to calving may reduce the incidence of milk fever. It also pays to add vitamin D to a high-phosphorus supplement fed to cows prior to calving whenever there is a risk of milk fever (Table 7.1).

Vitamin E

Vitamin E is a generic name for compounds called *tocopherols* which, together with selenium, regulate oxidative processes especially in nerve and muscle membranes. Grass and other green foods are rich in tocopherols, which are reasonably well preserved in silage but almost entirely lost during hay-making. The cereal germ, and by-products containing the germ, are also rich in vitamin E. Deficiency conditions such as white muscle disease were considered under selenium. Vitamin E is stored in the body and suspected deficiency states can be controlled by an injection usually containing both tocopherol and a slow-release source of selenium.

Last words on minerals and vitamins

Minerals and vitamins are essential elements of any diet. Moreover, a specific deficiency of a mineral such as selenium, needed only in minute amounts, may be sufficient to impair the performance of a grazing ruminant since metabolism can only march as fast as its slowest member. In practice, however, minerals and vitamins only assume importance in circumstances where the natural foods for the cow are deficient. This should always be considered a strong possibility when cattle are getting all, or nearly all, their nutrients from fresh or conserved grass. When dairy cows are fed substantial amounts of balanced concentrate supplements to promote high yields, mineral and vitamin deficiencies are rare. Feeding problems and metabolic diseases are far more likely to be due to shortages (or excesses) in the major elements of the diet needed to meet energy and protein requirement. It is dangerously easy to clutch at a symptom that has once, by someone, been linked to a deficiency of a specific mineral or vitamin and so prescribe a simple (and relatively cheap) form of therapy. It also tends to be relatively harmless but it should not be allowed to divert one's mind from more realistic (albeit expensive) sources of the problem.

Part III

Housing, Health
and Management

8 Healthy and Humane Housing

The basic environmental requirements to ensure satisfactory production, health and welfare in the dairy cow were defined in Chapter 4. To recapitulate briefly, these are:

(1) Thermal and physical comfort.
(2) Hygiene.
(3) Behavioural satisfaction.
(4) Optimal productivity.

These principles apply to all accommodation for cattle, which includes not only winter houses for lactating cows or growing heifers, but lairage at markets and slaughter houses. In designing accommodation for the dairy cow it is necessary to achieve the best possible compromise between her needs and the farmer's need for buildings and structures that are reasonably priced and convenient to maintain and operate. This chapter considers buildings and structures, so far as possible, from the point of view of the dairy cow. Those seeking greater details on structures, fixtures and fittings may consult Clarke (1980), Noton (1982), or specialist publications from the Ministry of Agriculture or Scottish Farm Buildings Investigation Unit.

Accommodation

In the UK dairy cows are housed for about seven months per year and spend five summer months at grass. This system of management is really only appropriate to cool, temperate maritime climates in which cows can get easy access to high-quality grass during the grazing season close enough to the milking machinery to permit twice-daily milking.

In hotter (or colder), drier continental climates it may be necessary to accommodate cows close to the milking parlour and bring food to them throughout lactation. The exact design of cattle accommodation is therefore influenced greatly by climate and by general management and feeding strategies. However, the general objectives are the same in all cases. The accommodation should be such that each cow within the herd has a clean, comfortable bed, reasonable shelter from the worst of the weather, sufficient access to clean, wholesome food and water, sufficient space to move around without difficulty or interference from other cows, and sufficient space and opportunity to express essential patterns of behaviour such as 'bulling' (oestrous behaviour). It should also be possible to direct cows without difficulty to specialist areas such as the milking parlour (of course), calving boxes, and restraining stalls for artificial insemination, veterinary inspection or treatment of common conditions such as lameness.

The cowshed

Since the cow can achieve thermal comfort over a wide range of air temperatures (see Fig. 4.1) she needs protection only from wind and rain or snow in the cold, wet areas of the world. In all but the most extreme hot, dry areas she needs protection only from the direct rays of the sun. To achieve physical comfort and to minimise the risk of infectious conditions such as environmental mastitis she needs a reasonably yielding, dry, clean bed. When moving between bed, feed, and milking parlour (etc.) she needs to walk over a surface that is not slippery nor a morass of mud of slurry.

In hot, dry climates it is possible to create simple shelters for dairy cows (see Fig. 4.2(b)) where they may rest under cover and out of the sun. Within this lying area, which should be bedded up with a material such as straw, each cow should be provided with sufficient lying space. The values in Table 8.1 of 3.2 to 5.8m² bedded area for cows ranging from 350 to 700 kg body weight should be taken as minimum standards. When sunshades can be erected cheaply from local materials, the covered bedding area per cow could be doubled to allow her to stretch out at a reasonable distance from her neighbours and so dissipate as much body heat as possible. When the climate is dry for most of the year the feed fence and 'loafing' area of the 'dry lot' need be neither covered nor surfaced with concrete. This also saves money and permits one to be generous with space, which can be very necessary when the rains come.

In the cool, temperate, maritime regions where cows are normally only confined during the winter months, the main problem is rain, not

Table 8.1   Dimensions and space requirements of cows and heifers

	Cows			Heifers, Friesian
	Jersey	Friesian	Holstein	2 yo
Body weight (kg)	350	600	700	450
Height to withers (m)	1.15	1.35	1.50	1.25
Body length (m)*	1.40	1.62	1.72	1.45
Reach of mouth (m)†				
at floor level	0.85	0.90	0.92	0.84
300 mm above floor level	1.00	1.05	1.07	1.02
Cubicle dimensions (m)				
length to wall	2.00	2.20	2.40	2.40
length behind trough†	1.40	1.60	1.80	1.60
width between partitions	1.10	1.15	1.20	1.15
height of neck rail‡	1.00	1.05	1.10	1.05
Feeding face, width (m)	0.55	0.70	0.70	
Loose housing (m^2)				
bedded area/head	3.2	5.0	5.8	4.0
feeding, etc./head	1.3	1.8	2.0	1.5

* tailhead to shoulders † = where present ‡ see Fig. 8.1

so much for its direct chilling effect but because any uncovered area of bare ground will rapidly degenerate into a sea of mud. In these circumstances it is essential for the cows to have a covered bedding area and for the walkways between bed, manger and parlour to be made of an impervious material that drains freely and can be scraped clean (i.e. concrete). Since roofed buildings and concrete floors are both expensive, the amount of space allocated to each cow is reduced to a minimum and 'loafing' space tends to disappear altogether.

There are three ways of accommodating dairy cows under cover:

(1) The cow house or byre in which cows may remain tethered in individual bedded stalls all winter and be fed and milked where they stand.
(2) A cattle court which is usually divided into a straw-covered area wherein cows may choose their own bed, and an unbedded, concrete, scraped area.
(3) The free stall or cubicle house in which cows are unconstrained and enter the cubicle of their choice to lie and rest.

The cow byre in which cows are tied up all winter has gone out of favour for a number of good reasons. Milking in the byre made sense when a bevy of beautiful milkmaids came to the cows, but once milking machines had become established it was more convenient,

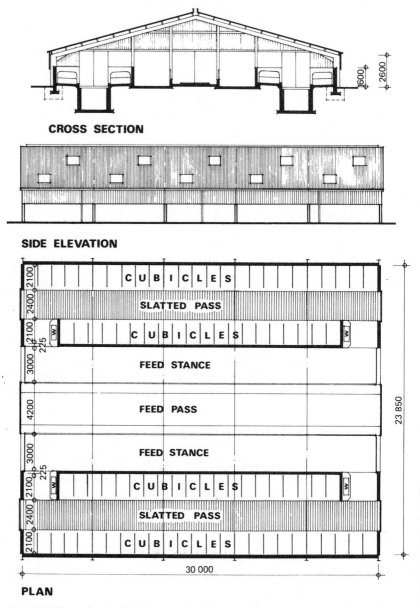

Fig. 8.1 Plan of a cubicle house for 90 cows (from Scottish Farm Buildings Investigation Unit 1985).

cheaper and more hygienic to bring the cows to the machines. The cow too has benefited from the demise of the old tie-stall. Even when she was generously bedded, the cow tethered all winter was prone to develop swollen knees and hocks and was especially susceptible to ketosis.

The current options are therefore for cubicles or a bedded cattle court. Although the recommended minimum bedded area for a Friesian cow in a straw yard is 5.0 m² compared with 2.5 m² for a single cubicle (Table 8.1), the amount of space allocated to passageways (etc.) in a cubicle house (Fig. 8.1) is such that the total covered floor space is similar in both systems – about 8 m² per Friesian cow. The Scottish Farm Buildings Investigation Unit in 1985 costed a cubicle house (with slatted passageways) at £106 per m² and a bedded cattle court at £84 per m². However, in a cubicle house a farmer could bed down four cows for the winter per tonne of straw. In a cattle court he would need about 1 tonne of straw per cow. The cattle court in which cows luxuriate in a deep, dry, clean bed of straw is ideal but only an economic proposition on farms which grow sufficient cereal to produce 1 tonne of straw per dairy cow. The dairy cow lying on a wet, dirty bed made from inadequate straw would be better off in a cubicle.

The problems of winter-housing dairy cows have been exacerbated by the switch from hay to silage as the main source of winter fodder. Silage is indisputably more nutritious than hay but it is very much wetter, so cows eating silage excrete far greater quantities of water in faeces and urine. This makes it more difficult to avoid wet bedding in a cattle court or large amounts of wet slurry in the passageways of a cubicle house. Both these things predispose to foot damage. Moreover, large amounts of slurry or foul water lying around in a cow house ensure that the bedding is seldom, if ever, dry and this encourages the growth of bacteria such as *Escherichia coli* which are responsible for environmental mastitis.

It is important therefore to minimise the amount of water lying within the cow house by arranging that passageways are slatted (Fig. 8.1) or scraped at least twice daily. It also helps to be aware that the average Friesian cow evaporates some 15–20 litres of water per day from the skin and respiratory tract. During a damp British winter the relative humidity in cow houses is often close to 100% – they are very clammy places indeed. The only practical way of removing water vapour from a cowshed is by natural ventilation. This is best achieved by the use of space boarding along the side walls to a depth of at least 1 metre below the eaves and space boarding from eaves to ridge at each gable end. Space boarding is made by fixing wooden planks (100 mm, or 4 inch) vertically with gaps of 20 mm ($\frac{3}{4}''$) between each plank. This permits air movement at low velocity but is extremely effective at keeping draughts, rain and snow out of the building. It also has the merit of letting in a lot of light. If the building has a central feeding passage then the roof ridge should also be open (Fig. 8.1).

Several other important features are also illustrated in Fig. 8.1 and Table 8.1. Since the width of cubicle plus stanchions is almost twice the space required for each cow at the feel fence, it follows that two rows of cubicles should be behind each feed fence. Water troughs are placed at each end of the inner row of cubicles and there is a passageway out from the cubicle area to the feed area at each end. This makes it easier for each cow to get where she wants to go and prevents a bullied or timid cow becoming trapped at the dead end of a passage by a real or imagined threat from another cow. The building in Fig. 8.1 is also generously equipped with transparent roof lights which, together with the space boarding, give good visibility even on sunless winter days. This makes inspection easier and is good for the morale of the stockman and (probably) his cows.

A variation on the cubicle house is the *cow kennel* which is a simple covered shelter wherein the sides of each cubicle are used to support the roof (Fig. 8.2). The relative economic merits of a cubicle house or

Fig. 8.2 Cow kennels.

kennels on a particular farm are outside the scope of this book. From the cow's point of view they are much the same.

Cubicle design

A cow cubicle or free-stall differs from a tie-stall in a traditional cow byre in that the cow may enter and leave at will and she gets her food somewhere else (Fig. 8.3).

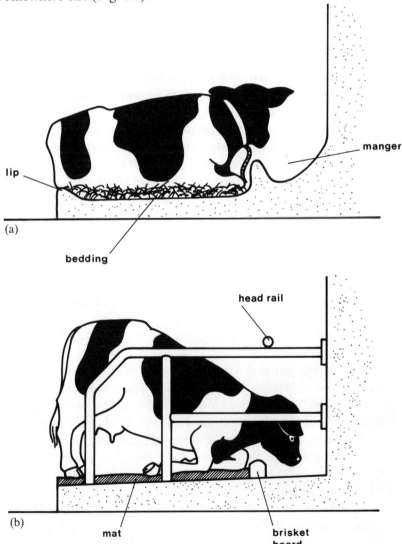

Fig. 8.3 Stalls for dairy cattle. (a) A cow lying in a tethered stall. (b) A cow rising to her feet in a free-stall or cubicle.

The specifications for a cubicle or free-stall for a dairy cow are that:

(1) The bed should be comfortable, clean, dry and free from contamination with faeces or urine.
(2) The cow should be able to lie down in a normal position and without risk of being trodden on or kicked by other cows.
(3) The cow should be able to stand with all four feet in the dry cubicle (not with the hind feet in the dunging passage).
(4) The cow should be able to change position from standing to lying (and vice versa) in a normal manner and without difficulty.

Considerable thought has been given to meeting these objectives since the first cubicles were put into practice in England by Howell Evans on his Cheshire farm in 1961.

First of all, the cow has to stand and lie with her head at the front of the cubicle so that her faeces and urine fall into the drainage passage. She is persuaded to enter the cubicle head first by providing a step up of about 250 mm from passage to cubicle. Cows are apparently reluctant to back into a cubicle up a step of this height, which has the second advantage of keeping out slurry from the passageways even when they are being scraped. The cow (nearly) always passes urine and faeces when standing up. The cubicle should be so designed that the cow stands with her back legs just in the cubicle but urinates and defecates in the passageway.

The cow normally lies down front legs first and stands up front legs last. During both movements, but especially when standing up, she needs to lunge forward with her head in order to maintain her balance (Fig. 8.3(b)). There should be sufficient space at the front of the cubicle to allow for this normal forward lunging motion. Some cattle in cubicles that are too short learn to change position in reverse order (i.e. get up front legs first), and even to adopt a 'dog-sitting' posture on their rumps with their front legs straight. This form of adaptation shown by some cows is no reason for building smaller cubicles.

When the cow lies down all her body should be on the bed. Her hocks, and especially her udder, should not overlap the lip at the rear end of the cubicle.

Recommended dimensions for cubicles are given in Table 8.1. Writing, as always, from the cow's point of view, I would say that a cubicle cannot be too long provided that the floor is well drained, the headrail (Fig. 8.3(b)) is adjusted to prevent the cow standing too far forward, and the stockman removes any excreta from the back of the cubicle beds twice daily. Cubicles that are too short predispose to mastitis, teat injuries and lameness.

The divisions between cubicles have evolved in response to experience and to fashion and there is now a bewildering array of more or less successful designs. The objectives are to align cows properly in their own cubicle, to prevent their feet interfering with or injuring their neighbours and to minimise the risk of injury to limb or teat as the cow changes position. Many of the early cubicles had solid horizontal rails at an elevation of about 300 mm (one foot) from the ground under which cows could trap or even break their legs. The modern trend is to minimise the amount of pipework or other rigid material by which the cow might get trapped. One effective solution is to replace the lower horizontal bar by (for example) polypropylene rope which is flexible, yielding and which can, in an emergency, be cut.

The cubicle bed

A clean, dry cubicle bed can be maintained on a base of rammed chalk or soil with a lip at the back of the cubicle, or on a base of concrete or bitumen macadam with an 80 mm slope from front to back and no lip to impede drainage. Whether the base is chalk, soil or concrete there is no doubt whatsoever that, above it, the dairy cow will benefit from some form of bedding. Her size and shape are such that the pressure exerted on skin and bone at sites like the knee and hock, and the shear forces exerted when in motion, are far more severe than those which we would experience if lying on the same surface. Anyone who doubts the need for bedding should spend a night lying on concrete! Small, non-lactating beef cows with small udders can be housed satisfactorily (although not ideally) on concrete slats in areas such as the Islands of Orkney where straw is prohibitively expensive. However, slats are not an acceptable bed for any lactating cow, whether of beef or dairy type.

Several types of bedding can be used successfully in cubicles. Each has its pros and cons.

Straw is the cheapest bedding when produced on farm. Its value is greatly improved if it is chopped since this makes it bulkier and more absorbent and reduces straw usage per cow by about 25%. The bed should be made to a depth of at least 50 mm if the base of the cubicle is made of concrete. New straw should be provided three times a week and any straw fouled with urine or faeces raked into the dunging passage twice daily.

Wood shavings make an excellent hygienic bed although they tend to be expensive. Depth of bedding and frequency of application should be the same as with straw (50 mm and three times per week). Sawdust is more absorbent than wood shavings and extremely comfortable. In damp conditions, however, fouled sawdust may be a good medium for the growth of *E.coli* and so predispose to environmental mastitis.

Sand is extremely comfortable if the bed is deep enough. (Think about it sometime when lying on a beach.) This requires a depth of 50 mm at the very least and 80 mm would be better. A thin coat on wet concrete becomes as abrasive as sandpaper. Cubicles bedded with sand need a lip to keep the sand in. The bed should be cleaned and raked daily but new sand need only be added at intervals of two to three weeks. If kept clean, it is very hygienic and makes a thoroughly satisfactory bed provided the slurry system can handle it.

Cow mats tend to be expensive but, as we saw in Chapter 4, extremely popular with cows. Within the range currently on sale, the mats with the most 'give' appear to be the most comfortable, for obvious reasons. Cow mats are potentially very hygienic, but the temptation to keep them spotless by liberal use of the hose should be resisted since in damp winter conditions they don't dry and can become slippery. It is better to keep the mats as dry as possible and cover them with a thin layer of bedding. A dry. interior-sprung cow mat, covered with a little clean, chopped straw, is just about the best possible bed for a dairy cow.

Passageways

The floor surface of passageways needs to be skid-resistant so that cows may walk freely without risk or fear of slipping and injuring themselves. On the other hand the surface should not be so rough as to cause excessive abrasion to the hoof and so predispose to foot damage. Finally, the surfaces should be as clean and dry as is reasonably possible since a build-up of wet slurry can soften and erode the hoof and so predispose to foot lameness. Moreover, if cows enter their cubicles with their feet soaked in slurry, this fouls the cubicle bed and increases the risk of environmental mastitis.

In Fig. 8.1 the floors of the passageways are made of concrete slats 125 mm wide with a 35 mm gap between the slats. This is a relatively expensive option but one which ensures that passageways stay clean, dry and free from slimy – and therefore slippery – moulds. New, solid concrete floors are tamped or brushed to roughen the surface slightly. Old concrete floors which have worn smooth and slippery can be restored by cutting shallow grooves with a special grooving machine. Newly-laid concrete floors in cow houses sometimes appear to increase the incidence of foot lameness not just because of the physical act of abrasion but also, I think, because wet slurry on new concrete can have a chemical, caustic effect which tends to erode the hoof.

If solid concrete is used the passageways should be scraped at least twice daily to avoid the build-up of slurry.

Calving and isolation boxes

Any dairy unit requires accommodation for cows that need special attention because they are sick or about to calve – in effect, the bovine equivalent of hospital beds.

The number of calving boxes required per 100 cows depends on how tight is the calving pattern (i.e. how short the period during which most cows calve) and how many cows are likely to calve while still out at grass. A clean pasture is a splendid, hygienic place for a cow to deliver a calf in the normal way but it presents problems if she has difficulties during calving and even worse problems if she cannot get up afterwards. The economics of the dairy industry in the UK encourage calving in the months of October to December, which implies that most cows will calve indoors. In these circumstances it would need five boxes per 100 cows to ensure first that each cow has access to a clean, disinfected box at calving and second, that any sick cow can be nursed in specialist accommodation for as long as necessary.

The specialist requirements for calving and isolation boxes are as follows:

(1) They should be separate from the main cow house although close enough to permit a cow that is sick, or about to calve, to be moved in with as little fuss as possible.
(2) Each box should be individually drained. The floors and walls should be rendered so that they can be power-washed or steam-cleaned.
(3) The boxes should be large enough for a calving cow to lunge about without risk of injury to herself, stockman or vet. This requires a floor area of at least 4×4 m. There should be at least one tethering ring, and preferably more, to restrain the cow's head (for example)

during assisted calving whichever way she has elected to lie down.

(4) There should be good artificial lighting and a protected electrical socket close to hand.

(5) The boxes and their doors should be sited so that any cow that dies can conveniently be pulled out by a tractor.

It should go without saying (but doesn't) that the needs of the cow for comfort and hygiene become even more paramount when she is sick or about to calve. Thus she needs a spotlessly clean, disinfected box with a generous, deep bed that will need to be mucked out completely when she leaves. In this case, unchopped straw is probably the bedding of choice on grounds of comfort and cost.

The most interesting approach that I have seen to the problem of accommodating cows at calving was on a very highly mechanised farm housing about 900 cows in cubicle units. After drying off, i.e. 6–8 weeks before they were due to calve, cows were moved to a maternity suite well away from the hurly-burly of the main dairy units. At first they were accommodated in kennels but, about two weeks before calving, were moved into deep-strawed cattle courts in groups of about 12 cows. Sufficient boxes were provided to accommodate each cow individually at the time of calving and these were equipped with a closed-circuit television so that the stockman could observe the cows when they calved, as expected, through the night.

I first saw this unit when it had been in operation for about three years. All aspects were working splendidly – except the closed-circuit television which had been abandoned after about six months. Before the erection of the maternity suite, the cows had been exposed to the noise and distraction of the main dairy units right up to the point of calving. In these circumstancess about 80% of cows calved between 10.30 pm and 6.00 am. In the peace and quiet of the new unit this pattern was reversed; nearly all cows calved between first and last visits at 6.00 am and 10.30 pm respectively. Needless to say the stockman, who didn't care much for television even when it showed pictures of cows, was delighted.

Heifer yards

The problems of housing dairy calves during their first winter are mostly concerned with minimising the incidence of infectious disease, especially pneumonia. I have dealt with this subject in detail elsewhere (Webster, 1985).

During their second winter, heifers should be housed in groups well matched for size and preferably in accommodation similar to that which they are likely to experience as milch cows. Thus, if they are to spend the winters of their adult life in cubicles they should gain experience of them before calving to minimise the confusion and stress that they experience when entering the adult world of cubicles, milking parlour and boss cows with a tetchy sense of who goes where and when. Dimensions for cubicles for heifers are given in Table 8.1. However, I repeat that from a cow's point of view a cubicle cannot be too long. Heifers can be discouraged from defecating in cubicles by placing the head rail further back. However, they are sometimes deterred from entering a cubicle by the presence of the head rail and may stand for long hours or, even worse, lie in the dunging passage. The good stockman allows his heifers to adapt to cubicles gently even if it does mean shovelling a bit more dung into passageways. Head rails can be adjusted when the heifers are ready.

The covered straw-bedded cattle court has one great advantage over the cubicle house for heifers in their second winter: it is much easier to see heifers bulling and so diagnose oestrus. Moreover, there is less risk of the heifers slipping and injuring themselves. If a bull, such as a Hereford, is running with the heifers, a straw yard is much better and safer for him too.

Handling facilities

Any dairy unit requires handling facilities which include:

(1) A race leading to a holding pen or crush in which most or all of the herd can be submitted to routine procedures such as blood sampling, tuberculin testing, vaccinations, etc. The race should also contain a foot bath. Most milking parlours are unsuitable for any procedures other than milking.
(2) Treatment stalls in which small numbers of cows may be held for artificial insemination (AI), pregnancy testing, foot treatment (Fig. 8.4), etc.

Marshalling areas should be designed to ensure that cattle are not forced to enter darkened areas or disturbed by sudden changes in floor surface (gullies, gratings, etc.). Because their centre of gravity is quite far forward they are uncertain when walking downhill. If this is essential it is better to use widely placed steps than ramps. Ideally they

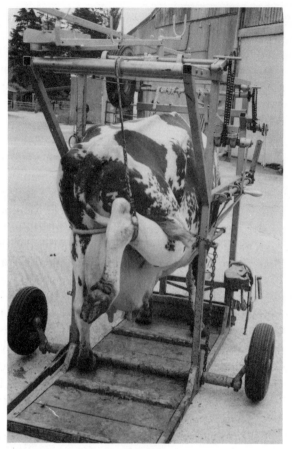

Fig. 8.4 Treating a cow's foot.

should be able to see where they are going but not distracted by things happening to left and right of them.

Cattle, like all ruminants, have a clearly defined 'flight zone'. So long as a potential threat such as a predator or man is perceived as being outside this flight zone, cattle will go about their business undisturbed. If man enters the flight zone slowly and calmly , and if the cow is free to move away, she will, equally calmly, do so. If man intrudes too far or too fast, or if she can see no way of escape, then she is liable to panic. There are many ways of designing good facilities for moving cattle (Grandin, 1980; MAFF Booklet 2495) and they are all based on this understanding of cow behaviour. So too is good stockmanship.

Individual stalls for cows awaiting AI should be sited under cover, close to the milking parlour and side by side so that cows have the reassurance of each other's company. Since one of the main roles of these stalls will be to accommodate cows while their feet are being trimmed or treated, it is necessary to have a bar to which a rope can be attached to lift, especially, a hind foot for treatment (Fig. 8.4). Front feet, which need attention less often, are probably best treated via a strategically placed side panel in a crush.

Lameness

The dictionary definition of lameness is a disability in movement of the limbs. The cow may be reluctant to move because it is painful, her movement may be impaired by paralysis, muscle weakness or joint damage, or she may suffer a combination of all these things. It is important to remember that lameness may be confused with a disorder totally remote from the feet or limbs such as pleurisy or traumatic reticulitis (Chapter 10). Both hurt the cow when she moves.

It is, however, fair to say that 90% of cases of lameness in cows are due to inflammation or injury to the hoof or skin between the hooves. Other afflictions of the limbs include swollen knees and hocks caused by bad lying conditions. On the whole these do not seem to cause a great deal of disability unless the skin is broken and a deep-seated infection develops. Occasionally a cow may dislocate her hip or fracture her pelvis through slipping and falling, especially when weakened by calving and/or hypocalcaemia. Amazingly these cows often manage to struggle on and even milk quite satisfactorily but they lose condition and it may prove as economic as it is humane to send them for slaughter at once. If they are in too much pain to be moved they should be slaughtered on the spot. I shall return to this subject later.

My main concern, however, is with foot lameness since this is probably the single greatest insult to the welfare of the modern dairy cow. Surveys of foot lameness based on cases treated by veterinary surgeons indicate an average annual incidence of about 5–6%. When cases treated by farmers are included, the incidence rises to about 25% per year. Inspection of the feet of cull cows at slaughter reveals that nearly every animal is suffering or has suffered some form of foot damage. In other words, the dairy industry is living with a painful, crippling disease with a morbidity rate close to 100%!

To understand foot lameness, it is necessary to consider first the structure of the cow's foot (Fig. 8.5). The hoof of each claw is made up of two parts, the *wall* which grows down from the coronet, and the *sole*.

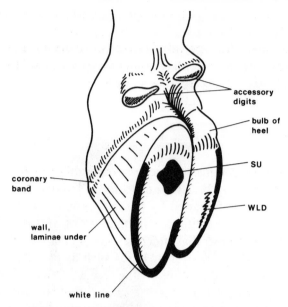

Fig. 8.5 Structure of the cow's foot. SU = solar ulcer, WLD = white line disease.

The junction between the wall and the sole is called the *white line*, which is confusing because in cattle is isn't white (it is rather whiter in the horse's hoof, which is where the name originates). The horny, insensitive tissue of the wall and sole are joined to the living, sensitive tissues of the foot by the laminated folds of the *corium* which create the area of contact between the insensitive hoof (which may be thought of as a shoe) and the sensitive tissues of the foot. The laminae of the corium also act as shock absorbers, spreading the impact when the foot makes contact with the ground over a surface area far greater than that of the sole of the foot.

It is necessary, first, to identify the most common sites of injury and disease in the foot and then consider how they arise.

White line disease

This is a disintegration and separation of the junction between the sole and the wall of the hoof (Fig. 8.5). It exposes the living tissues to infection which may track up the laminae of the wall or across the sole, causing inflammation with the formation of pus. The pus breaks out to the surface either at the coronet or, having completely underrun the sole, through the softer skin at the bulb of the heel. While the pus is trapped within the hoof the pressure and pain are intense.

Solar ulcer

This is a discrete ulcer appearing typically in dairy cows at the junction between the sole and heel on the outer claw of the hind foot (Fig. 8.6). The cow is likely to become lame before the site actually ruptures. Inspection at this time will reveal a thin, soft, bruised area of sole that

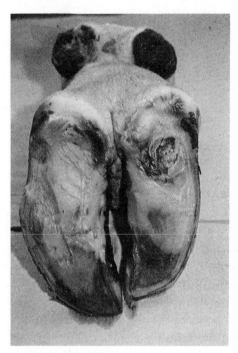

Fig. 8.6 A solar ulcer.

is painful to the touch. Once open, the ulcer is a constant source of infection but pus can escape. What usually happens in untreated cases is that granulation tissue ('proud flesh') grows out through the holes. At this stage the condition appears to be at its most painful.

'Foul-of-the-foot'

'Foul-of-the-foot', or interdigital necrobacillosis, is the most common of a range of country names for an infection due to *Fusobacterium necrophorum* which arises usually in the interdigital space as a complication of a wound from a stone, flint or other sharp object. The whole foot becomes very hot and swollen which has the effect of separating the claws. If the foot is washed and scrubbed it is usually possible to find the initial wound and diagnose *Fusobacterium necrophorum* infection by the presence of pus and a characteristic foul smell, which is conspicuously worse than the normal smell of a cow's foot.

Aseptic laminitis ('founder')

Laminitis describes any form of inflammation of the highly vascular laminated corium that links the hoof to the rest of the foot. Entry of infection following a primary separation of wall and sole at the 'white line' will lead to a septic laminitis with the formation of pus. Acute aseptic laminitis arises in the previously undamaged foot and in the absence of infection when inflammatory substances are released into the blood supply to the laminae as a result of some primary disturbance to digestion or metabolism. The condition is commonly known as 'founder' in both cattle and horses. The skin above all four feet is usually hot to the touch although the hooves themselves may feel cold. If the cow has to stand, she tends to put her weight onto the bulbs of the hocks to relieve the pain on the laminae, adopt strange postures to relieve the pressure on the worst affected feet, and to 'paddle', especially with her hind legs, for the same reason. In many cases she will lie down and show extreme reluctance to rise.

To appreciate the pain of laminitis it helps to contemplate crushing all our fingertips in a door and then attempting to walk around on them. The short-term complications of laminitis are considered below. The cow with chronic laminitis develops distinct horizontal ridges across the wall of the hoof and the toes become markedly overgrown to give a slipper-like appearance (Fig. 8.7) due, in part, to their habit of standing on their heels.

Baggot (1982) and Blowey (1985) described in more detail these and other much less common afflictions of the foot, like sandcracks. Blowey also describes in detail the treatment of the affected foot. My main concern is prevention based on an understanding of the factors that cause, or predispose to, foot lameness.

Genetics

The conformation of the modern dairy cow undoubtedly predisposes to foot damage. Over 70% of all causes of foot damage occur in the

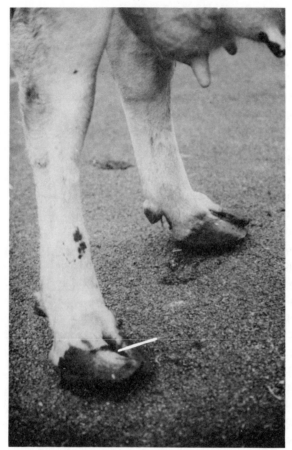

Fig. 8.7 Chronic laminitis – a slipper foot.

hind feet of dairy cows and, of these, over 70% involve the outer claw. In bulls, on the other hand, foot lameness tends, if anything, to be more prevalent in the front feet. It is impossible to escape the conclusion that the abnormally large udder of the modern dairy cow has distorted the posture and gait of her hind limbs in such a way as to predispose to foot damage. Prior to calving the cow puts *less* weight on the outer claw. This may lead to overgrowth of soft horn. The expansion of the udder at calving alters the alignment of the hind legs throwing *more* weight onto the outer claw (Greenhalgh, MacCallum and Weaver, 1981). A second genetic element in the aetiology of lameness is that black hooves are harder than white. It is therefore reasonable to assume that black hooves in an otherwise desirable Friesian or Holstein bull would be something of a bonus.

Nutrition

In the UK, the three most common clinical signs of lameness in high-yielding dairy cows – aseptic laminitis, white line disease and solar ulcers – are typically seen in the early winter, shortly after cows have (a) been housed and (b) given birth. On these grounds we should be suspicious both of housing design and feeding in early lactation. A closer look at the evidence suggests that while winter housing is a contributory factor, most cases of lameness are concentrated within the first 60 days after calving irrespective of the interval between time of housing and time of calving. Cows calving at pasture in spring are less likely to be affected.

I have stressed the point more than once that the primary nutritional cause of aseptic laminitis is ruminal acidosis, usually provoked by excess intake of starchy foods. It is possible that high protein in concentrate rations can exacerbate the problem, due perhaps to absorption of vasoactive peptides. We do not yet know exactly what vasoactive substances are directly responsible for laminitis nor where they are released. However, it does seem that high protein concentrates do not cause laminitis unless the cow also eats sufficient quickly fermentable carbohydrate (starches and sugars) to cause ruminal acidosis. Starchy, cereal-based concentrates should therefore be introduced gradually, starting preferably before calving. No more than 4 kg concentrate should be given at a single meal to adult Friesian cows and no more than 3 kg per meal to heifers. If this requires out-of-parlour feeding then so be it.

Aseptic laminitis in early lactation may occasionally be provoked by a generalised toxaemia resulting from mastitis or metritis (infammation of the uterus, Chapter 12) but this is a much rarer cause than improper feeding of starchy concentrates. One of the main reasons why cows calving on spring pastures are less prone to laminitis is that they eat less cake.

Acute laminitis not only causes severe pain but tends to separate the horn of the hoof from the sensitive layers underneath. This makes the white line more vulnerable to separation as the result of a skid or stepping on a sharp object. Separation of the laminae, plus the fact that the cow is standing on her heels, also causes the hoof to rotate forward, thereby further separating the heel end of the sole from its blood supply. The horn becomes soft, new growth is inhibited and it takes very little abrasion from the ground to create a solar ulcer. Improper feeding is therefore a major contributory factor not only to aseptic laminitis but also to white line disease and solar ulcers – in short, nearly all cases of lameness in housed cows.

Environment

'Foul-in-the-foot' or interdigital necrobacillosis can usually be attributed to a wound inflicted by a sharp object in filthy conditions such as occurs when cows walk through a muddy farm gate into which bricks, rubble, etc., have been tipped to prevent them from sinking out of sight. It is therefore most frequent in cows at pasture in wet conditions on smaller, poorer, less productive farms. Control is simple in theory: make the walkways cleaner and safer. In practice, however, this may be prohibitively expensive for the small farmer.

Damage to the sole of the foot, whether primary or secondary to acute laminitis, is predominantly an affliction of housed cows. The problem does seem to have become worse since the first cubicle houses came into operation 25 years ago but this does not necessarily imply that the cubicle house is inherently worse than the tie-byre or straw-bedded cattle court. In the same period cows have (until very recently) been fed more and more concentrates and silage has largely replaced hay as the main winter forage. The problem of starchy concentrates has been considered already. Silage, although a far better food than hay, does generate large quantities of very wet faeces. Wet slurry certainly softens and probably erodes the hoof and so predisposes to excessive wear and a decreased resistance to shear forces. Once again the solution is to keep the passageways as clean and dry as possible.

Behaviour

Cows that were tethered all winter in old-fashioned byres hardly moved at all so that, whereas they were more prone to afflictions of joints, they were less likely to injure their feet. The wall of the cow's hoof grows at a rate that is particularly well suited to a life that involves a modest amount of exercise on moderately hard ground. The cow in a cubicle house gets insufficient exercise to ensure even wear on the feet but does move around often enough to run a reasonably high risk of injury. White line disease, although facilitated by improper nutrition, is usually precipitated by a skid or a stone.

Foot lameness is a particularly severe and distressing problem in heifers calving for the first time. While this can often be attributed in large part to improper nutrition, it can also be caused, indirectly, by the psychological problems faced by the heifer when she joins the adults in the main cow house. Uncertain what to do and submissively giving way to the older and more dominant cows in the herd, she may fail to achieve her proper period of rest in a cubicle but stand for excessively long periods in the dunging passages. If these are wet they will soften the feet. Even if they are dry, the heifer who comes off pasture and then stands for as much as 20 hours on concrete is likely to become very foot-sore. More specifically, there will be erosion and bruising of the sole. Such an animal needs only a slight degree of nutritional laminitis to provoke a solar ulcer.

Foot care

Control of foot lameness based on what has been discussed already may be summarised as follows:

(1) Gradual introduction to starchy concentrates, starting before calving, and not too much at any one meal.
(2) Non-skid passageways.
(3) Introduction of heifers to the routine of cow house (and milking parlour) in time to ensure that they have adapted fully before they calve down for the first time.

Lameness can also be kept under control by foot bathing and foot trimming. The most commonly used solution in a footbath is 5% formalin which not only acts as a disinfectant but also tends to harden the hoof. Formalin can operate in extremely dirty conditions but ideally the cows should be directed to pass through two footbaths, the first containing water simply to wash the feet, and the second the 5% formalin. Cows should enter the foot bath once a week and be made to stand in the diluted formalin solution for one minute. Cows' feet need to be inspected regularly and trimmed as necessary in a standing like that illustrated in Fig. 8.4. Blowey (1985) describes the process of foot trimming very well but really it is a practice best learnt by experience under the direction of a skilled operator such as those who run courses for the UK Agriculture Training Board. In essence, the intention is to keep healthy feet the right shape by removing overgrowths of wall and toe. Treatment of a seriously infected foot is a matter for the veterinary surgeon who will administer an appropriate antibiotic, remove sufficient damaged horn to allow free drainage of pus and protect the affected area from contact with the ground and risk of re-infection by applying a bandage and possibly a boot. In very severe cases it may be necessary to amputate one claw. Most cows recover well after the amputation of a single claw and can continue to lead a useful life.

Transport and slaughter

Cows need to be transported safely and humanely from farm to farm and through markets and, at the end of their lives, deserve a decent death. We must recognise that they are timid creatures who are alarmed, at least initially, by any novelty that may appear to constitute a threat. They are reassured by the normal routines of a working day and by the presence of other, familiar members of the herd. The physical and social environment of a cow in transit or lairage should therefore be as close as possible to that to which she is accustomed. It would, however, be anthropomorphic in the extreme to presume that a cow was under stress simply because she was in a lorry that was 'taking her away from home' or that, in a well-run slaughter house, her passage down the race to the stunning pen was inherently more stressful than her passage down a race to a tuberculin test. Each procedure needs to be considered simply in terms of the immediate problems faced by the cow.

Loading and unloading

Walking up a ramp into a lorry is a potentially alarming experience for a cow that is not used to it. Cattle that have been transported frequently usually enter lorries much more readily, which strongly suggests that they do not remember transport as a particularly alarming experience. The main problems of loading and unloading *for the cow* are:

(1) the fear and risk of injury when walking up and especially down steep slopes.
(2) fear of entering an unknown, dark area.
(3) alarm created by sudden noise.

To avoid these stresses the slopes on loading ramps should not exceed 20° and the sides should, of course, be protected, preferably with solid partitions. Widely placed steps with a tread of 300 mm and a step of 100 mm are better than sloping ramps. If ramps are used, the floor should be designed so that the cow's foot cannot slip far *in any direction*. Bars running horizontally across a ramp will not stop the foot slipping sideways. The entire surface of the ramp should be non-skid. One approach is to create 25 cm squares using weldmesh rods, with the bottom rods running up and down the ramp and the upper rods across it. This makes cleaning easier.

Aluminium is tending to replace wood in cattle lorries since it is longer lasting and easier to clean and maintain. However, aluminium, or any other metal ramps, are very noisy compared with wood and should be covered by a non-slip matting or generous amounts of straw. The inside of the vehicle should be well-lit so that the cow can see where she is going. She should not be hurried but allowed to proceed at her own pace.

After leaving school I worked on a dairy farm where lived an old Shorthorn-type cow called Bluebell whose performance was dismal but whose disposition was phlegmatic. Whenever any adult cattle were to be transported, Bluebell was kept in with them. She was haltered and led into the lorry first to encourage the others. When all were aboard she was led out again. Fortunately for my peace of mind, I left the farm before she did.

Unloading from a lorry is potentially easier because cattle are moving from darkness into light. It is, however, particularly difficult for cattle to descend a steep slope. The off-loading area at markets, lairages and, wherever possible, on farms, should have a ramp or step so that the animals can leave the lorry via the ramp/backboard down as gentle a slope as possible.

The journey

The potential problems of the journey may be assessed within the context of the 'five freedoms' outlined in Chapter 1. Problems of the journey arising from these are:

(1) Excessive deprivation of food and water.
(2) Heat stress, cold stress and physical discomfort leading, at worst, to physical exhaustion.
(3) Injury from fittings within the vehicle, or other cattle; subsequent disease following infection *en route*.
(4) Aggressive encounters with other cattle.

The European Convention for the Protection of Animals states that animals shall not be left more than 24 hours without being fed and watered. This period should only be extended if the journey can be completed in reasonable time. In other words, if the journey can be completed within (say) 26 hours it would be an absurdity to submit cattle to the stress of unloading and reloading to offer them food and water when only 2 hours short of their destination.

Lactating cows should be milked twice daily. Dry cows in good condition tolerate 24 hours without food and water with little, if any, distress because the amount of undigested material in the rumen will continue to yield a substantial flow of nutrients for at least this time. In fact it may pay to restrict the food intake of cull cows to hay only prior to transport, to restrict the output of wet faeces and so improve comfort and hygiene within the vehicle.

The thermal environment within cattle vehicles is usually controlled (somewhat) by natural ventilation through spaces in the walls. When adult cattle are being transported it should be possible to adjust the openings to provide from 20% to 40% open area according to conditions. The lower value applies in very cold conditions and when the vehicle is nearly empty. When a vehicle full of cattle is kept stationary for any period of time, as many ventilators as possible should be open. This implies that the driver must, of necessity, be a good stockman who can recognise a hot or cold cow when he sees one. The driver is also primarily responsible for ensuring that cows experience as little physical discomfort as possible during the journey and so arrive at their destination in a fresh condition. Space allowances for adult cows should range from 1.1 to 1.8 m^2 per cow for animals weighing 300–700 kg. It is recommended that not more than five adult cows or eight young heifers should be penned together on lorries. On railway vehicles the number of adult cattle should not exceed 10–12. When the driver proceeds smoothly on good roads, cattle will stand in any direction and often lie down after 1–2 hours. The rougher the drive the more likely they are to remain standing.

Although it is common for cows from several farms to travel on the same lorry, they should be kept apart to prevent aggression or the threat of aggression that inevitably accompanies the meeting of strangers. Horned cattle should not travel with those without horns. Herd bulls travel best in the company of two or three old, familiar cows, not in oestrus.

Killing with kindness

The decision to kill a cow may be taken because she has become more valuable as a red meat animal than as a continued producer of milk or because she is suffering from a disease or injury from which she is unlikely to recover or which would be too expensive to treat.

Let us consider first the slaughter of cows that are healthy and fit to travel to an abattoir to be killed for human consumption. The animals should all be fit at the outset of the journey. However, they should be inspected on arrival at the abattoir. Any cow that is injured and incapable of walking without undue pain should be killed in the vehicle. The environment in lairage should meet the same minimum requirements as on farm. The cows should be able without difficulty to stand up, lie down and move around to food and water when provided. The journey to the stunning pen should be calm, quiet and unhurried, ideally through a curved, solid-sided race sufficiently narrow to prevent a cow turning round but able to admit an attendant in the event of an emergency. The curve in the race prevents cows from seeing what is happening at the end of the line.

The purpose of stunning is to render an animal instantaneously unconscious prior to sticking and bleeding out. The procedures most commonly used for stunning cattle, the captive bolt or percussion stunner, are probably the best methods in existence for rendering any animal instantaneously unconscious. The most important welfare requirement for adult cattle is that they are then bled to the point of death as quickly as possible so that no animal recovers any degree of consciousness. There is a belief in the meat trade, based mostly on myth, legend and the Old Testament, that it is necessary for the heart to beat during bleeding out to remove 'all' blood from the meat. Recent research has shown this to be an unfounded belief which means that killing and bleeding out do not have to be done at the same time.

The UK Farm Animal Welfare Council has concluded that bleeding an animal without prior stunning, as in the Halal or Shekita methods of religious slaughter, is an unacceptable practice. Those who defend these practices claim that a single cut with a sharp knife is not painful. This may be so, but it is no defence of religious slaughter. Many things in a cow's life (and ours) may cause more acute pain. What is totally unacceptable is the distressing fact, for the cow, that she is conscious of choking to death in her own blood.

For most dairy farmers the greatest welfare concern relates to casualty and emergency slaughter. Casualty animals (in England) are those which are not in severe pain but which are deemed to be suffering from sickness or injury sufficient to require slaughter as soon as possible. It is both economically expedient and humane to transport such animals reasonably short distances to a local abattoir but unrealistic to expect it to stay open for business 24 hours a day, seven days a week.

Emergency cases may be defined as those where the animal has experienced severe injury and cannot move or be moved without incurring severe pain. Here, the Farm Animal Welfare Council recommends that animals should be killed humanely on the spot. If all or any of their carcass is to be passed for human consumption it must arrive at the slaughterhouse accompanied by a veterinary certificate indicating that the carcass is 'unlikely to have been rendered unfit for human consuimption'.

The Council of Europe requires that an animal be eviscerated within 45 minutes of death. In the UK, evisceration should be carried out 'within a reasonable period'. The farmer who can find a friendly neighbourhood slaughterhouse which will accept a dead cow with a veterinary certificate in an emergency, can call on his vet to kill and bleed the animal where it lies. He who cannot – and as abattoirs get bigger and grow further apart, this number is likely to increase – may only be able to get knacker prices for a dead cow and is faced with a very difficult conflict between economics and humanity which may be expressed, 'does my charity and compassion to one old cow amount to (say) £300 in cash which I can ill afford to lose?'. In this, as in all welfare problems, it is important to be as sympathetic to the needs of the farmer as to those of his animals. One solution to this problem would be a compulsory (but low premium) insurance scheme operated by an organisation such as the Milk Marketing Board to compensate the farmer if, in the interests of humanity, he arranged for such animals, after veterinary inspection, to be slaughtered on the spot and taken away by the knacker man.

9 Milking and Mastitis

A typical good beef cow feeding her own calf may produce 10 litres milk per day which will be consumed by the calf in about seven serious meals. The single calf drinks from only one teat at a time, may or may not drink from all teats, removes perhaps 1.5 litres at a meal and almost certainly leaves some milk within the sinuses of the mammary gland. A typical good dairy cow being milked twice daily at intervals of 10 and 14 hours may yield 36 litres/day in lots of 15 and 21 litres per milking. All four quarters are milked simultaneously until the flow of milk has practically ceased.

The two procedures are very different.

When I first worked on a dairy farm in the 1950s, each cow had her last drops of milk 'stripped' by hand milking after the clusters were removed. When I asked different people why, I was told variously that milk left in the udder either caused the cow to dry up or was likely to cause mastitis. I remember offering the suggestion then that this didn't seem to be a problem for beef cows and was told by Fred Robinson (to whom this book is dedicated), 'an' a beef cow don't give too much milk does'n?', which shut me up for 30 years, i.e. until now, when I have attempted to analyse the argument from first principles.

Figure 9.1 (taken from Theil and Dodd, 1977) illustrates the changes that take place in udder pressure (kPa) and milk excretion rate (kg/h) in the hours after milking. At first, pressure in the milk sinuses is atmospheric and secretion of milk from the alveoli is maximal (here 1.4 kg/h). There is a small, initial rise in pressure in the sinuses then little further increase in pressure for another 8 hours. After 10 hours without milking, pressure starts to increase steeply and when it reaches 3 kPa at about 12 hours, milk secretion rate starts to fall. At 32 hours intramammary pressure is at a peak of 8 kPa and milk secretion ceases. The message of Fig. 9.1 is that twice-daily milking is compatible with maximum rates of milk secretion for most cows provided that the intervals between milkings are evenly spaced. The more the milking interval is unbalanced from 12:12 towards the extreme of 8:16, the

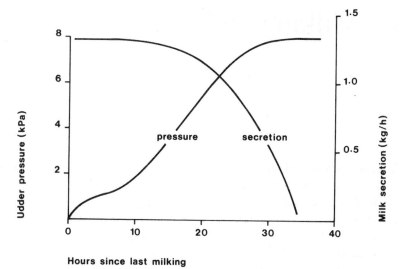

Fig. 9.1 Changes in intramammary pressure and milk secretion rate with time after milking.

more discomfort the cow will experience initially and the less milk she will produce.

Experiments where cows have deliberately been milked to only 75% of capacity for 20 days have shown that this chronically inhibits milk secretion. Moreover beef cows do produce more milk if stimulated to do so by being suckled by two calves. However, the small amount of residual milk or 'strippings' left in the udder after conventional milking (about 0.5 kg) is not sufficient to affect the total lactation yield of the cow. Thus the old practice of hand stripping was wasted effort since it did not increase overall milk yield, yet indirectly, it may have been associated with a lower incidence of mastitis than the much more damaging procedure of leaving the milking machine on the cow until the last drop of milk had been removed.

Following the natural stimulus from her calf or conditioned stimuli such as entry to the milking parlour, feeding and udder washing, the cow secretes oxytocin which contracts the myoepithelial cells around the alveoli and so encourages the 'let down' of milk into the mammary sinuses (see Fig. 3.5). Thereafter the rate at which milk is released is determined mainly by the diameter of the streak canal. The effect of the oxytocin-mediated milk ejection reflex lasts only a few minutes so that if the period between let-down and onset of milking is delayed or if the duration of milking is unnecessarily prolonged it will outlast the period of let-down and more milk will be retained in the udder. Thus a hand-milker, however skilled, is unlikely to extract as much milk as a machine milking all four teats at once. Moreover a cow who is an abnormally slow milker by virtue of narrow streak canals is also likely to give less milk overall.

Some dairy farmers who have the labour available, milk their highest yielding cows three times daily. This is sound in principle since it reduces pressure on the udder and can, in the right circumstances, lead to increases in yield of the order of 10%. It can, however, be a fairly exhausting practice for the cows as well as for the milkers if, in addition to getting milked three times daily, they have to wait on each occasion for perhaps 90 minutes in collecting yards. This can seriously erode the time available for the essential business of eating and resting and so wipe out the advantage of more frequent milking unless the queueing time involved with each milking session can be reduced to a minimum.

The use of computer-directed robotics has revolutionised manufacturing industries like the motor trade which used to depend on men endlessly repeating routine tasks. Computer-controlled feeders that dispense food to cows on demand are already commonplace in the dairy industry. There is no reason, in principle, why it should not be possible to design a computer-controlled milking parlour with feeding facilities that a cow could enter whenever she felt hungry or in need of milking, there be fed an appropriate amount of concentrate and be milked by a computer-driven robot. Automatic cluster removal is available now. Automating the process of preparing the teats and applying the clusters should present no insuperable problems. The machine would, however, have to recognise the possibility of mastitis and, if in doubt, call for help from the herdsman.

The labour-saving aspects of such a robot are obvious. What is even more attractive is that if it worked properly and was meticulously backed up by the service of a good herdsman/technician, it would probably be beneficial to the cow in terms of production, health and welfare, increasing yield, relieving discomfort from a grossly swollen udder, and perhaps thereby reducing the incidence of both mastitis and lameness.

The milking machine

The first attempts to introduce a mechanical substitute for the chores of hand-milking involved the insertion of metal tubes through the teat canal to allow milk to escape by force of gravity and intramammary pressure. Given the rudimentary understanding of hygiene in the early 19th century this must have been fraught with danger and it is no surprise that the method never caught on. The first effective machines to apply a vacuum to the teat and remove milk by suction were developed in the 1850s. The modern design of milking machine which uses a pulsator to create an intermittent vacuum at the teat and so mimic the sucking action of the calf, was patented by one Dr Shiels of Glasgow in 1895. The vacuum applied to the teat pulsated rhythmically between 50 and 15 kPa. Although there have been enormous developments to the mechanics of the milking machine, the pulsation and the applied vacuum are essentially the same now as those proposed by Dr Shiels.

Figures 9.2 and 9.3 illustrate the workings of the milking machine in just sufficient detail to explain its effect on the cow. The mechanics of plumbing, operation, cleansing, etc., are described in detail in the publication *Machine Milking* (1977, Eds. Theil and Dodd). In essence, a constant vacuum of 40–50 kPa (12–15 inches, or 300–380 mm of mercury) is applied to the inside of the teat cup liner. The pulsator transmits an intermittent pulsator vacuum (PV in Fig. 9.3) to the space between the rim of the teat cup and the liner. When pulsator vacuum is equal to the vacuum inside the liner and thus in the milkline (45 kPa in Fig. 9.3), the vacuum within the liner sucks milk from the teat. When the pulsator vacuum is removed and pressure between teat cup and liner returns to atmospheric, the constant vacuum in the milk liner causes the liner to contract round the teat, squeezing and interrupting the flow of milk.

The greater the vacuum applied to the milk line the more rapid the flow of milk, so the less time taken to milk each cow. However, the lower the vacuum the gentler the action of the teat cups and the lower the risk of injury or mastitis. The optimal vacuum as assessed by the vacuum gauge in a particular parlour is further affected by the siting of the gauge and idiosyncracies of the plumbing, and needs to be determined and set by an expert.

The change in PV cannot be instantaneous but should be as fast as possible, i.e. a and c in Fig. 9.3 should be as brief as possible. The overall *pulsation ratio* (%) is given by:

$$100 \ (a+b)/(a+b+c+d)$$

and should normally be close to 60%.

The flow of air and milk in bucket machines and pipeline machines in cowshed and parlour is illustrated in Fig. 9.2. Typical values (in

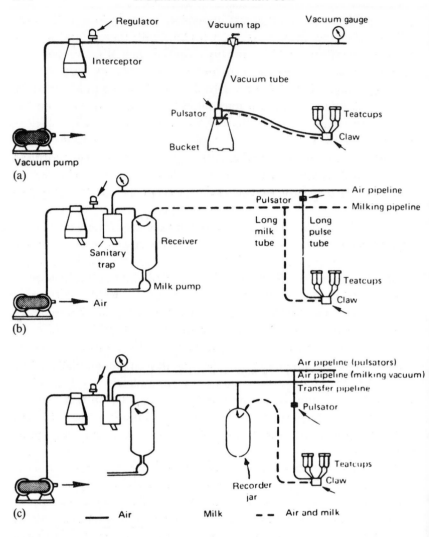

Fig. 9.2 Flow of air and milk in milking systems. (a) Bucket machine. (b) Milking pipeline machine. (c) Recorder machine (from Theil & Dodd 1977).

fact, our own) for pulsation frequency of 55/min and pulsation ratio of 65% (which corresponds to an open ratio of 58%) are given in Fig. 9.3, which also illustrates the effects of the milking machine on the teat during the period of maximum milk flow and during the last phase of milking, or 'machine stripping', when the rate of milk flow has

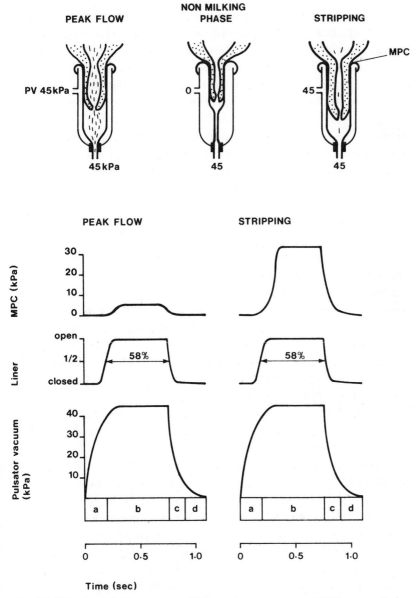

Fig. 9.3 Function of the teat cup. PV = pulsator vacuum, MPC = mouthpiece chamber. (For further explanation, see text.)

decreased to about 10% of peak. During the period of peak milk flow the potential vacuum of 45 kPa applied to the teat extracts milk from the teat at about 20 ml/sec into the short milk line to the recorder jar. This rapid flow of milk ensures that the actual vacuum applied to the teat within the mouthpiece chamber (MPC) of the teat liner is quite small (5 kPa in Fig. 9.3). However, during the stripping phase, the flow of milk is insufficient to offset the effect of the vacuum in the milk line and vacuum within the MPC increases to 30 kPa or more. The effect of this increase in suction is to draw the teat deeper into the liner, or conversely to allow the liner to 'crawl' up the teat. When milk flow ceases altogether the mechanical effect of the liner on the teat increases further. This explains why overmilking is one of the main factors predisposing to teat injuries such as prolapse of the teat sphincter or 'black spot', a granulating sore on the end of the treat.

Teat damage incurred during milking and the effects of milking on a previously damaged teat obviously increase the risk of bacteria entering a quarter through the teat canal and inducing mastitis.

Another feature of many current milking machines that may predispose to mastitis is the phenomenon called *teat-end impact*. During the non-milking phase, milk previously drawn from all four quarters lies within the central claw of the cluster that joins the short milk lines from each teat cup to the long line to the recorder jar. When the pulsator vacuum is reapplied to the liner, the first effect is to suck milk back into the teat cups from the claw. Especially during the stripping period, when the teat has penetrated deep into the liner, there is a considerable vacuum at the MPC and within the teat canal. Mixed milk from the claw, which may contain bacteria from an infected quarter, can strike the teat end with some force and be sucked right up through the streak canal into the teat itself and so transmit infection from one quarter to another.

This problem can be reduced but not eliminated by the use of moulded teat cup liners with a shield at the bottom that disperses the impact of milk flowing backwards from the claw and prevents it striking the teat sphincter directly. We use shielded lines and are very pleased with them. Transmission of infection from one quarter to another via refluxed milk can be avoided altogether by the inclusion of gravity-operated ball valves within the claw that prevent any return of milk to the short milk tube. This approach holds great promise but still (as I write) requires some development. The claws are very heavy and MPC vacuum shifts are not yet entirely satisfactory.

Teat damage and teat-end impacts can be caused by fluctuations in vacuum pressure in the short milk tubes (for example) when clusters are being put on another cow or when a cow kicks its clusters off. Once again, the design of a milking unit to minimise local fluctuations in vacuum pressure is a job for an expert. Teat-end impacts also occur when air is allowed to enter the space between one teat and the liner. This can occur if a cow dislodges the cluster (perhaps because her teats are sore) or if one or more cups slip down the teat. The noise caused by the inrush of air is unmistakable. Entry of air down one teat cup increases the amount of milk refluxing into the other cups whose vacuum seal is still intact, and so increases the risk of injury and infection. For the same reason dairymen are instructed to (but don't always) turn off the vacuum before removing the cluster at the end of milking. One of the advantages of a system that includes automatic cluster removal is that it doesn't take short cuts.

The milking parlour

Most developments in parlour design have been directed towards decreasing capital and labour costs and thus fall outside the scope of this book. It seems to me that the conventional herringbone parlour with a central pit that places the cows' udders at a comfortable working level suits both cows and dairymen very well. There was some years ago a craze for rotary parlours conceived by engineers who thought it would increase efficiency if machines circulated the cows on a carousel to a stationery operator. Rotary parlours became a popular buy for research institutes whose directors were more impressed than farmers by the 'high-tech' nature of the equipment and less aware than farmers of two conceptual weaknesses in the design:

(1) the inherent unreliability of moving parts, especially in the cool, wet conditions of a milking parlour, and
(2) the fact that whereas the design permits all clusters to be donned at a single spot, it also ensures that any excrement is spattered over the widest possible area.

So far as the cows are concerned, the design of the milking parlour is immaterial, provided that:

(1) They can enter and leave with minimum fuss, without slipping or having to make excessively sharp turns in narrow passageways, and without interference from other cows.
(2) They can stay in parlour long enough to consume their concentrate ration. Most Friesian cows given 4 kg would finish eating within 8 minutes. The parlour routine should give them at least this long.
(3) The proper hygienic procedures are followed (see next section).

(4) The clusters are not left on too long. Most cows milk out within 4–5 minutes and it is obvious when they have finished. The most common reason for overmilking is that the milker is under too much pressure and cannot remove the clusters as soon as each cow is ready. Here again, automatic cluster removal is a good idea so long as it works. A first-class milker operating a routine that gives him sufficient time to remove the clusters from all cows as soon as they are ready is still the safest option.

Hygiene in the milking parlour

Meticulous attention to hygiene in the milking parlour is essential to ensure the production of clean, wholesome milk and to minimise the distress and loss of production associated with mastitis. This means doing everything possible to prevent bacteria entering the udder or the bulk milk tank. Before any cow is machine-milked her teats and the adjacent area of her udder need to be washed and the milk from each quarter needs to be inspected for signs of mastitis. The sequence of events here is important. A sample of milk from each quarter should be drawn by hand onto the shiny black plate of a strip cup and examined for abnormalities such as clots, flakes, pus, watery discolouration or streaks of blood. These can be seen best before 'let-down' so this examination should take place before the udder is washed.

In our parlour at the University of Bristol Farm the herdsman first briefly rinses the teats of the cows as they arrive using a warm water hose (40°C), then hand-milks samples for examination, then washes the teats and adjacent region of the udder using the warm water hose and his hands only. This suffices for most cows but if a cow has managed to get her udder particularly filthy it is sprayed more thoroughly with the pressure hose used to wash down the parlour. Cloths are not used to wipe the udder. If a warm water hose is not available and the udder must be washed down using a bucket of warm water and a cloth, the water should contain disinfectant approved for dairy use (usually based on chlorine or iodine) and individual cloths should be used for each cow, since the use of a common cloth can transfer infection from one cow to another. Individual paper towels are excellent for drying the udder after it has been washed in running water from a hose, but are unsuitable for washing. The alternative, which is to provide boiled individual cloths for each cow, is horribly arduous. It sounds dangerous but it is true to say that it is better not to wash the teats at all than to use a common udder cloth. Fresh, warm running water from a hose is undoubtedly best of all.

At the end of milking the teats and udder need to be protected from the conditions the cow will meet when leaving the parlour, which may, in winter, be a cold, damp, relatively dirty cubicle house. Disinfection of teats after milking can be achieved by dipping or spraying them with an approved disinfectant (there are many) and an emollient such as glycerol or lanolin which protects the damp skin of the teats from chapping or cracking when the cow goes out into the cold air.

When cows leave the milking parlour their teat sphincters are still relatively relaxed. If they immediately went back to lie down on a cubicle bed contaminated with *Escherichia coli* it would be relatively easy for these organisms to gain entry. Dipping or spraying teats after milking may accelerate the closure of the teat sphincters but it is also good practice to prevent cows from returning to their cubicles for 30 minutes after milking. If they can wait under cover and with access to food and water, so much the better.

Total bacterial counts

The Milk Marketing Board encourages the production of clean milk by paying a premium or imposing a penalty according to the total bacterial count (TBC) in the bulk milk sample. Since 1982 a premium has been paid only to those farmers whose milk contains no more than 20,000 microorganisms per ml (which sounds a lot but isn't). Bulk samples with TBCs over 100,000 per ml will incur a penalty, which is doubled for persistent offenders. High TBCs can usually be attributed to improper cleaning of the milk lines or incorrect control of temperature in the bulk tank, both of which allow proliferation of microorganisms originating from outside the udders. However, high TBCs can also be due to milk leaving the cows with a high bacterial content. The good herdsman should have identified any cow with clinical signs of mastitis, probably treated her at once with an intramammary antibiotic prescribed by his veterinary surgeon, milked her separately into a 'dump' line and thoroughly washed out the lines and short milk tubes before re-use on healthy cows. However, as we shall see, a number of cows with chronic mastitis can appear clinically normal to even the best herdsman. Their milk will enter the bulk tank and elevate TBC. The incentives to reduce TBC are thus in the interests of all concerned parties – the consumer who deserves wholesome milk, the producer, and the cow herself since it should reduce her risk of getting mastitis.

Mastitis

Mastitis is an infection of the mammary gland that produces clinically recognisable signs of pain and inflammation of the udder, contaminates the milk and may produce toxins which cause generalised sickness and even death. This general definition covers a range of diseases of differing severity caused by at least 40 different microorganisms, the most common being *Staphylococcus aureus, Streptococcus agalactiae, Escherichia coli* and *Corynebacterium pyogenes.* The verterinary surgeon needs to know the specific organism responsible for a particular case of mastitis in order to ensure the most effective treatment. For the farmer however, it is more useful to recognise mastitis as three distinct disease complexes, each with its own different natural history.

(1) *Contagious mastitis.* This describes a group of infections caused by bacteria which can survive and multiply on the skin of the teat and in the udder and which are therefore transmitted from cow to cow during milking. The bacteria most commonly isolated from cases of contagious mastitis are *Staphylococcus aureus* and *Streptococcus agalactiae.*

(2) *Environmental mastitis.* This describes mastitis caused by organisms such as *Escherichia coli* which do not normally live on the skin or in the udder but which can enter the teat canal when (for example) the cow lies down on a contaminated cubicle bed or in a calving box.

(3) *Summer mastitis.* This condition occurs typically in dry cows and in heifers at pasture on warm, humid summer days. The organism most often associated with summer mastitis is *Corynebacterium pyogenes* which is transmitted by females of the head fly *Hydrotaea irritans.*

Surveys of mastitis in the UK indicate that an average herd of 100 cows would expect about 40 cases of mastitis per year. Reports of over 100 cases per 100 cows per year are not uncommon and very few farmers achieve an annual incidence much lower than 30%. The average annual incidence does seem to have declined slightly in recent years, possibly as a result of the policy to minimise TBC and mastitis cell counts in milk, but there has been a substantial change in pattern. Fifteen years ago the great majority were cases of contagious mastitis transmitted from cow to cow in the milking parlour and environmental mastitis was relatively rare (about 5% of all cows). Today environmental mastitis is the most prevalent type, accounting for about one-third of all cases in the UK. This suggest that whereas the management of contagious mastitis has got better, the management of environmental mastitis has got worse.

Contagious mastitis

Forty years ago when penicillin was just coming into veterinary practice, the most common cause of contagious mastitis was *Strep. agalactiae*. This organism grows and reproduces only within the udder so that the only source of new infection is the chronically infected carrier cow. The only time that transmission can occur is during milking when bacteria are transmitted from quarter to quarter in infected teat liners and encouraged to enter the teat canal by mechanical factors predisposing to teat-end impact.

Infection with *Strep. agalactiae* in a previously clean quarter causes acute inflammation with pain, heat and swelling. The foremilk usually contains large clots. The cow may have a slightly elevated temperature and be depressed and off her food for a couple of days. The organism is sensitive to penicillin and treated cows return apparently to normal within a few days. However, the infection may persist in the quarter in a sub-clinical form. The pain and swelling go and the milk appears normal although there may still be a few clots in the foremilk. Within the quarter, however, a low-grade chronic infection may be steadily destroying the secretory cells and especially the ducts, which are replaced by scar tissue, so reducing the capacity of the quarter to produce milk. The cow with a chronically infected quarter is not only less and less productive but is a constant source of infection to other quarters and other cows.

Even if the first attack of *Strep. agalactiae* mastitis is completely cured, re-infection is always likely while there are any infected animals in the herd. Subsequent infections cause less acute inflammation and pain and may go unnoticed but continue to erode the milking capacity of the quarter. Because *Strep. agalactiae* only lives within the mammary gland and because it has remained sensitive to penicillin it is relatively easy to control. Its place as public enemy number one in the arena of contagious mastitis has been usurped by a much more difficult organism, *Staph. aureus.*

Staph. aureus is a natural inhabitant of the skin of cows and man where it does no harm unless the cow's teat or the milker's hand is cut or cracked, when *Staph. aureus* can cause the wound to turn septic. If the organism is able to penetrate the teat in sufficient numbers the disease takes one of two forms. *Peracute* staphylococcal mastitis can occur, rarely, but especially in early lactation when the defences of the mammary gland to infection are at their lowest. The cow rapidly becomes very ill with a high fever and severe depression and may become comatose and die within 24 hours of the first signs of disease. The infected quarter is grossly swollen and extremely sore, which makes the cow very reluctant to move. The secretion from the quarter is usually a blood-stained serous fluid. If the cow survives, blue, gangrenous patches appear on the quarter and progress to blackened, oozing sores. Although the cow with peracute *Staph. aureus* can be saved by antibiotics if caught in time, the quarter is almost invariably lost. It is fortunate that this condition is rare since it is catastrophic.

The much more common form of *Staph. aureus* mastitis is less severe but *chronic.* The cow may not appear ill and the infected quarter may not be very painful. The foremilk may or may not show clots and wateryness according to the severity of the condition at the time. However, like *Strep. agalactiae, Staph. aureus* slowly but surely destroys the normal tissues of the quarter leaving it more and more scarred and steadily reducing milk output.

Treatment is complicated by the fact that there are many types of *Staph. aureus* with differing sensitivities to antibiotics. Moreover, it is practically impossible to eliminate the organism from an infected quarter during lactation even though the cow may appear clinically normal. Once again, therefore, the infected cow is not only steadily losing performance but is a threat to others.

The most effective form of treatment for *Staph. aureus* mastitis is *dry cow therapy.* A long-acting antibiotic, known to be effective against the strain of *Staph. aureus* which is known to be most prevalent in the herd, is introduced into the quarter using an intramammary tube when the cow is dried off. This is designed to persist in the quarter long enough to kill all the bacteria. However, if a cow has had several attacks of *Staph. aureus* mastitis in the same quarter (and if she is worth saving), a second dose of antibiotic should be introduced two to three weeks after the first.

Because it is ubiquitous, hard to eliminate from an infected quarter and may be resistant to many antibiotics, *Staph, aureus* is now the most important cause of contagious mastitis. The *control* of contagious mastitis (especially that caused by *Staph. aureus*) involves attention to all of the following details:

(1) Good hygiene in the milking parlour before, during and after milking.
(2) Regular checks on the milking machinery, correct vacuum and pulsation, clean, soft liners that have retained their original shape.
(3) Use of teat shields or claws with ball-valves to reduce or eliminate teat-end impacts.
(4) Early diagnosis and appropriate treatment of clinical cases.
(5) Bacteriological examination of milk from clinical cases and apparently normal cows to discover the most common sources of infection and their antibiotic sensitivity.
(6) Dry-cow therapy for all cows.
(7) Culling of cows that experience four or more attacks of contagious mastitis in one quarter in any one lactation.

Mastitis cell counts

The first line of defence in the udder is formed by *neutrophils* (Chapter 4). These increase in number in response to infection and help to effect a cure or at least bring the infection under control. These neutrophils plus epithelial cells, shed from the udder at a rate proportional to the amount of damage taking place at the time, enter the milk. A high concentration of these 'mastitis cells' per ml of milk sampled from one cow or one quarter is an excellent indicator of sub-clinical chronic mastitis. Average mastitis cell counts for the herd are measured by MMB on samples taken from the bulk tank. Absolute values and trends in cell counts can be used to indicate the prevalence of cows carrying contagious mastitis and the progress of a control scheme. In 1975 the average cell count for the UK dairy herd was 586,000/ml. By 1985 it had fallen to 393,000/ml with over 30% of herds below 300,000/ml. There is no doubt that this decline in cell count has been associated with a decline in *contagious* mastitis.

It has been argued that environmental mastitis has increased as contagious mastitis has declined and that since the main contributors to cell counts are neutrophils which are defenders rather than attackers, we may, by striving for low cell counts through culling and

dry cow therapy, be reducing the resistance of our cows to environmental mastitis. There is no good epidemiological evidence to support
this assertion. We may conclude that a high concentration of
neutrophils in the milk from a single quarter is evidence for a
chronically infected quarter and a high cell count in the bulk tank
evidence for a lot of chronically infected quarters; thus a herd with a
high cell count requires a systematic programme for the control of
contagious mastitis.

Environmental mastitis

As indicated earlier, this complex disease is caused by organisms which
are natural inhabitants of the microenvironment that surrounds the
cow and harmless in the right place. The most important organism is
E. coli which is a natural inhabitant of the large intestine and continues
to survive and grow wherever cows' faeces are permitted to lie. The
next most important is *Strep. uberis* which is a natural inhabitant of
cows' skin, mucous membranes and faeces. Infections with *Strep.
uberis* are usually less severe than those with *E. coli*, and are sensitive
to penicillin. The principles for control of both infections are
essentially the same so I shall concentrate on environmental mastitis
caused by *E. coli* organisms, hereafter called simply coli.
     There are hundreds of types of coli which differ in their sensitivity to
antibiotics and their capacity to release endotoxins which spread
through the body and produce a general sickness. The effect produced
by the entry of coli into a quarter varies greatly according to the state
of the immune defences of the cow, in particular her ability to recruit
neutrophils to repel the invasion. In late lactation or during the dry
period, any coli that enter are likely to be overwhelmed by neutrophils.
In late lactation the herdsman may observe signs of inflammation but
if he took a milk sample it could well be sterile because the neutrophils
had already destroyed all the invaders. In early lactation, especially
the first few days, the cow's capacity to mount a neutrophil defence is
greatly diminished. In these circumstances, local signs of infection may
be relatively minor (since inflammation is an expression of the body's
defence to infection) but the coli are able to multiply and liberate
endotoxin which makes the cow feel very ill indeed. Her temperature
is elevated, she shivers continuously, goes entirely off food, usually
develops a severe diarrhoea and so rapidly becomes dehydrated, with
a tight skin or 'hidebound' appearance and sunken eyes. The mortality
rate in such cases is high.

Severe, generalised toxaemia with little local inflammation in early lactation, and the complete localisation and rapid elimination of infection that occur in the dry cow, are opposite extremes of the contest between cow and coli. Most infections fall somewhere in between: the cow has local inflammation in the udder; within 24 hours or less the foremilk becomes a straw-coloured liquid with few, if any, clots; the cow shows some signs of toxaemia, being slightly fevered and reluctant to eat. This sort of mastitis is difficult to distinguish from contagious mastitis caused by *Staph. aureus* or a wide range of other organisms without taking a milk sample for bacteriological examination (which takes time), so the vet cannot guarantee to prescribe the most appropriate antibiotic at the first visit.

To *control* environmental mastitis, everything possible must be done to keep faeces and faecally contaminated bedding away from the teat sphincter, especially during the first 30 minutes after milking. Hygiene in the court or cubicle house was described in Chapter 8: dry, clean bedding, scraped passageways or slats, control of relative humidity. I wrote then and emphasise even more strongly now how important it is to keep calving boxes spotlessly clean. Entry of coli into the udder during calving is likely to cause the most severe form of *E. coli* mastitis with general toxaemia and a high risk of death. Farmers who, for economic reasons, operate a very tight calving pattern, i.e. calve most of their cows within a very short period, run a greatly increased risk of coli mastitis being contracted in the calving box since even steam-cleaning cannot guarantee total elimination of coli from a calving box and they should, after cleaning, be rested for at least one week and longer if at all possible.

Good hygiene in the milking parlour is also essential to minimise the risk of faecal contamination of the teats when the clusters go on. This is why it is strongly recommended that teats be dried with an individual paper towel after the initial udder wash since one drop of water contaminated with faeces can contain enough coli to cause mastitis.

The increased incidence of environmental mastitis in recent years can be attributed in part to damp, dirty conditions especially in cubicle houses where cows are eating large quantities of wet silage. The association between increasing environmental mastitis and diminishing mastitis cell counts is, as I wrote earlier, probably coincidental. However, coli mastitis does not normally proceed to a chronic phase with persistently high neutrophil counts. Thus a farmer may, by dint of a careful control programme for contagious mastitis, reduce his herd's cell count without necessarily affecting the incidence of environmental mastitis, although in most cases good hygiene in and out of the milking parlour will reduce the incidence of both disease complexes.

Another reason for the increase in environmental mastitis is that it is more common in the cows that give the most milk. Selection for high yielders, who are also fast milkers, inevitably selects for cows with wider bore teat sphincters which are more likely to admit coli and other organisms. There is little that one can do about this except try harder to achieve good hygiene.

Summer mastitis

This condition affects cows and heifers at pasture during the summer months. It nearly always involves dry cows, partly because so many cows in the UK calve in the autumn. Several different organisms can usually be isolated from the infected quarter but the most important is *C. pyogenes* which is transmitted from cow to cow by the head fly *Hydrotaea irritans*. These flies land on all parts of the cow but are particularly attracted to secretions such as blood, tears and sweat. A slight cut or sore on the end of a teat will attract the fly and allow bacteria to enter the udder.

Most cases of summer mastitis are acute and severe. The cow shows all the classic signs of a generalised toxaemia: fever, depression, weakness and inappetence. The udder is hard, swollen and painful. Many cows die during this early acute stage and others abort. Those who survive develop massive abscesses in the infected quarter. The strippings are thick, yellow and putrid smelling, bearing a grotesque resemblance to rotten custard. The affected quarter is nearly always destroyed for ever.

Summer mastitis can take a milder form which may escape notice in a dry cow at pasture but which can cause sufficient scarring in the sinuses and teat canal to block the secretion of milk when the cow or heifer calves.

The following steps are necessary for the *control* of summer mastitis:

(1) Dry cow therapy using a long-acting antibiotic not only helps to rid the udder of chronic infections such as those caused by *Staph. aureus* but also radically reduces the risk of *C. pyogenes* infection during the dry period.

(2) Flies should be controlled using either a spray, which should concentrate on the areas around the udder and abdomen, or

insecticidal ear tags. The insecticide from these gets itself spread fairly evenly over the coat when the cow grooms herself although concentration on the skin over the udder tends to be relatively low. Spraying is therefore better if done regularly but insecticidal ear tags are effective in controlling summer mastitis and have the further welfare benefit that they reduce irritation about the cows' heads.

(3) Dry cows and heifers should, as far as possible, be grazed away from woods and standing water during the months of July to September (in UK).

(4) Be meticulous in shepherding dry cows at pasture to ensure that any affected animal is treated without delay.

Antibiotic residues in milk

Milk marketing boards rightly impose strong sanctions against farmers who allow the milk they sell to become contaminated by antibiotics. The sensitivity of the current test for the presence of 'inhibitory substances' in milk corresponds to 0.006 parts per million of penicillin. The penalty for a first failure in 1985 was £90, rising to £149 for third and subsequent failures. The most common reasons for failure can be attributed to incompetence or deliberate neglect of veterinary advice, i.e. accidental or deliberate failure to withhold milk from treated cows for the required period. Failures that occur on farms that have obeyed all the rules can usually be attributed to unforeseen prolonged excretion of antibiotic from cows treated during lactation (which is bad luck), or an insufficient lapse of time between dry cow therapy and calving. Different preparations for dry cow therapy have differing stated degrees of persistence, from 7 to 51 days. It is important to ensure that the appropriate preparation is used. This, like so much else, requires that farmer, herdsman and veterinary surgeon all have access to good, clear records for each cow.

Economics of mastitis

Blowey (1986) has calculated the costs of mastitis, varying in severity from the mildest form that requires only one course of intramammary tubes administered by the stockman (£17.40 at 1985 prices) through acute, severe mastitis requiring one visit from the vet and incurring chronic loss of production (£76.50) to the rare, fatal case of (for example) *E. coli* or summer mastitis (Table 9.1). Assuming that these are distributed 70:29:1, this gives an average cost of about £40 per case, or £2,000 per year for a 100 cow herd experiencing 50 cases per year.

Table 9.1 An assessment of the economic cost of mastitis (1985 values adapted from Blowey, 1986)

| | Severity of mastitis | | |
	Mild	Acute	Fatal
Probability per 100 cows	70	29	1
Cost (£/case)			
vet's fee	–	17.50	35.00
antibiotics etc.	2.40	10.00	23.40
loss of milk, discarded	15.00	21.00	4.50
chronic loss	–	27.00	–
loss of cow	–	–	500.00
Total cost	17.40	75.50	562.90

Blowey operates a specific mastitis control programme based on one or two preventive medicine visits per year. The average number of cases of mastitis in herds participating in his programme fell from 51/100 cases in 1980 to 32/100 cases in 1985 and the percentage of cases that recurred fell from 25 to 10. All the herds had been using most conventional aids to hygiene in the milking parlour before the start of this scheme so the only new cost was about £60 per annum for the preventive medicine visits, which involve routine checks on each aspect of control for contagious and environmental mastitis, examination of individual records of treatment for mastitis and subsequent decisions as to what further diagnostic tests or treatment may be necessary and what individual chronic cases may need to be culled. This annual expenditure of £60 has yielded an average increase in income of £720, which must be deemed a very satisfactory return.

10 Miscellaneous Maladies

An animal (any animal) suffers a malady when all or part of the body is physically damaged or destroyed or chemically poisoned by:

(1) physical injury;
(2) pathogenic micro-organisms or their toxins;
(3) deficiencies or excesses of normal dietary constituents or products of their digestion and metabolism;
(4) poisonous plants and chemicals.

The common infectious disorders of calves, enteritis and pneumonia, are caused by pathogenic micro-organisms (viruses and bacteria) that are endemic, i.e. always present. By the time heifers calve for the first time they have become immune to most of the endemic organisms that can enter the body via the gut or respiratory tract. Thus scours and pneumonia, the major hazards of young calves (see Webster, 1985), are seldom seen in adult cattle. The four major diseases complexes seen in the dairy cow all arise directly from the special stresses and challenges faced by the breeding, lactating female. These are:

(1) Infections of the mammary gland, i.e. mastitis.
(2) Disorders of the reproductive tract, e.g. metritis, infertility.
(3) Physical 'wear and tear', perhaps complicated by infection, e.g. lameness.
(4) Metabolic disorders, e.g. ketosis.

Three of these have been considered already within the context of the systems of feeding, housing and management from which they emerge. Afflictions of the reproductive tract leading to infertility will be covered in Chapter 12. The vast majority of health problems in adult cattle fall into one of these four categories.

A comprehensive textbook on bovine diseases (which this is not) would include a fifth category of disease caused by internal and external parasites which are, or may be, endemic to the environment of the dairy herd. I shall deal with these very briefly because they can usually be controlled quite simply and so should not constitute a severe threat in practice.

There remains a miscellany of maladies which are hard to categorise except in so far as they can be attributed to injuries, infections or combinations of the two, that occur sporadically, cannot be linked to systems of husbandry and management – and are, in short, due to bad luck. These range from infections produced by exotic micro-organisms such as foot-and-mouth disease to complete 'acts of God' such as being struck by lightning. I shall, in this chapter, discuss some general points concerning the recognition and treatment of disease in cattle and mention the more common of these maladies as they occur and are controlled in the UK. Readers seeking a comprehensive international review of diseases of cattle and their differential diagnosis should consult Blood and Radostits (1989).

Signs of disease

The best way to learn how to recognise signs of ill-health in a cow is to stand and stare at healthy cows 'long and long', for signs of ill health always appear as departures from normality in posture, movement, alertness, appetite, etc., and the first signs may be very very subtle. Thus the good, observant herdsman is likely to spot the onset of ill health *in his own cows* before the vet. He will be aware that something is wrong but may not know exactly what, nor therefore exactly what to do. The following section is intended to help the stockman interpret the early signs of ill health. In Tables 10.1 – 10.4 the more common causes of death and disease have been tabulated in relation to their first and secondary symptoms. These generate a list of primary suspects and so determine what action is most appropriate. The lists of suspects that appear in the tables are by no means comprehensive, nor is it suggested that they are a diagnosis, merely the first steps in that direction.

Sudden death

The stockman discovers one (or more) dead cows, usually at pasture. Although the death may appear sudden to the stockman, it may not have been so for the cow – an hour can be quite a long time when one is dying. He should now look for secondary symptoms. These include signs of struggling or convulsions before death, the extent of bloat, blood at the nose and anus, frothy discharge from the mouth and nose, and evidence of burns, especially around the nose and feet. Table 10.1 relates these symptoms to the primary suspects. A bloated carcass does not necessarily mean that bloat was the cause of death since all carcasses bloat after death. He should therefore consider other factors such as grazing history, state of the pasture and evidence of bloat in other cows.

Table 10.1   Possible causes of sudden death in adult cattle

First symptom	Secondary symptoms	Suspect
Sudden death	struggling ⎫ frothy nasal discharge ⎬	hypomagnesaemia
	extreme bloat ⎫ bleeding from orifices ⎬	primary bloat anthrax
	none —————— ⎰	plant poisons (e.g. yew) perforated ulcer peracute mastitis
	burns ——————	lightning

A cow found dead at pasture after a very wet night in the first week after turn-out in spring, with much froth about the mouth and nose and evidence of severe convulsions, almost certainly died of hypomagnesaemia. In this unusually straightforward case the preliminary diagnosis has been made on five different pieces of information, possibly reinforced by a history of inadequate magnesium supplements. Most cases are far more difficult. A small amount of blood loss from the nose, mouth and anus is not uncommon in cows that are killed by primary bloat but it can also be a symptom of anthrax – a notifiable disease communicable to man. For this reason, any cow found dead without obvious cause should be reported to the Police or direct to the Ministry of Agriculture so that a veterinary inspector can (at no charge) take a blood sample and examine it for the presence of *Bacillus anthracis*.

Anthrax is extremely rare. Other causes of sudden death listed in Table 10.1 such as lightning strike or poisoning by yew or laburnum are also very uncommon. Nevertheless, unless the cause of sudden death is obviously bloat or hypomagnesaemia, the possibility of anthrax should be eliminated before the cow is moved to a knacker's yard where a veterinary surgeon can conduct a proper post-mortem examination.

Drooling

Drooling, usually accompanied by other signs of discomfort about the mouth, can signify a range of maladies (Table 10.2). If only one cow

Table 10.2   Possible causes of drooling in adult cattle

First symptom	Secondary symptoms	Suspect
Drooling – one cow	choking swollen, painful tongue swollen jaw	physical obstruction wooden tongue tooth abscess lumpy jaw
Drooling – several cows	diarrhoea mouth ulcers	bovine virus diarrhoea
	nervous signs	poisons, e.g. rhododendron, organophosphorus compounds
	lameness mouth ulcers	foot & mouth disease

is affected she may have choked on a piece of a root crop or wood. If her tongue is swollen and painful, particularly at the back of the throat, she is probably suffering from 'wooden tongue', an infection caused by *Actinobacillus lignieresi*. If the jaw is swollen the primary cause may be a tooth abscess or the specific infection 'lumpy jaw', caused by *Actinomyces bovis*. In the unlikely event that one or more cows are found to have ulcers on the tongue or the mucous membrances inside the lips and cheeks, then suspect bovine virus diarrhoea or foot-and-mouth disease. (I shall return to these two later.) At this stage, it may be said that most cases of drooling in cattle are treatable but all except simple cases of choke require veterinary attention without delay.

Abdominal symptoms

Whereas severe bloat usually indicates a primary disorder of rumen fermentation or a complete obstruction of the oesophagus, some degree of bloat can accompany abomasal displacement (Chapter 7) or traumatic reticulitis ('hardware disease' or 'wire'). It is not uncommon for cows to eat sharp pieces of wire, nails, etc., which almost invariably settle in the reticulum. Usually these pieces of hardware do no harm but occasionally they may penetrate the reticulum wall, puncture the diaphragm, enter the chest and, at worst, puncture the heart. A cow that stands with her back arched, grunts with pain on the slightest movement, has a slightly elevated temperature and is (probably) slightly bloated, should be suspected of having traumatic reticulitis and veterinary attention should be sought. As a first aid measure, the cow should be tethered in such a way that she stands with her front feet higher than her hind feet, thereby relieving the pressure of the reticulum and the penetrating foreign body on the diaphragm.

Although the faeces of adult cows eating grass or silage are often very wet, they do not often suffer from diarrhoea which is a clinical sign of

Table 10.3 Diseases possibly associated with abdominal symptoms

First symptom	Secondary symptoms	Suspect
Bloat, severe	none	primary bloat
mild	drooling, choking	physical obstruction
	pain, grunting	traumatic reticulitis
	acidosis (laminitis)	primary nutritional upset
Diarrhoea, acute	fever, inflamed udder	toxaemic mastitis
	fever (abortion)	salmonellosis
chronic	weight loss	parasitism
		Johne's disease
	mouth ulcers, (abortion)	BVD, teart
Straining	constipation	intestinal obstruction
	nervous signs	ragwort poisoning
	vaginal discharge	calving or abortion urogenital tract infection

Secondary symptoms in parentheses occur only intermittently

malabsorption. Acute diarrhoea in a large number of cows can often be attributed to a primary nutritional disorder associated, for example, with a change of concentrate ration. Acute diarrhoea in a single cow associated with a high fever may be caused by salmonellosis or *E.coli* mastitis (Table 10.3). Chronic diarrhoea, accompanied by weight loss in large numbers of cows, is probably due to gastrointestinal parasitism. If so, it is really inexcusable. Scouring and weight loss in heifers could be due to teart (Chapter 7). Chronic diarrhoea in one or a few individuals could be caused by BVD or Johne's disease.

Abdominal straining is caused more often by discomfort in the urinary or reproductive tract than by problems in the gut (Table 10.3). Intestinal obstruction is very rare. Ragwort poisoning is a possibility but dairy farmers (surely!) recognise ragwort as a poisonous plant and ensure that it does not get into hay or silage cut in mid to late summer. Cows normally avoid ragwort in the field but may fail to recognise it or be unable to avoid it in conserved forage.

Nervous symptoms

The first symptom of many diseases is a change in the cow's state of mind. She may, often quite suddenly, become hyperexcitable: becoming wild-eyed, bellowing aimlessly and overreacting to the slightest stimulus. This is most often an early symptom of hypomagnesaemia although in recent years bovine spongiform encephalopathy (BSE) must be considered as a possibility. In some cases hyperexcitability may be caused by tetanus or one of several poisons.

Other diseases may cause the cow to appear disorientated to a degree that ranges from being mildly bemused to a state of delirium. Specific signs include head shaking, walking in circles, a partial or complete blindness and pressing the head against a wall or post. This last symptom has been variously interpreted but I think it indicates a severe headache. These signs can be caused by a wide range of conditions which include overeating of concentrates (in which the cow may appear drunk), some cases of ketosis, listeriosis which is an infectious encephalitis contracted, very rarely, by cows eating silage, lead poisoning and vitamin A deficiency. These last two conditions are more likely to occur in young calves or yearling heifers.

The nervous signs associated with BSE usually start out as mild and sporadic but become progressively severe as the cow's personality disintegrates. The first signs may be sudden angry kicks in the parlour or 'spooking' at unfamiliar objects and a general sense of apprehension. Later, the cow with BSE may stand unsteadily gazing fixedly at nothing in particular with eyes that are called wild by some but to me look simply bewildered.

A cow that is suffering from a systemic infectious disease is likely to be depressed but, in this case, there should be many, more specific, symptoms. Sometimes, however, cows appear depressed, perhaps with some loss of appetite, but there are no other obvious symptoms. It is difficult to know what is wrong in these cases.

It could be that she is suffering from some fatty degeneration of the liver and so is feeling somewhat 'liverish'. She could also be in the early stages of a large number of diseases which could include a first attack or relapse of hypocalcaemia (milk fever) or even botulism. I must stress that botulism, tetanus, listeriosis and other conditions listed in Table 10.4 are extremely rare. Readers scanning the tables of symptoms in this chapter should resist the temptation to develop a vicarious hypochondria with respect to the cows in their charge, similar to that of the neurotic reader of the *Home Doctor* who imagines him(her)self to have the symptoms of every disease in the book.

The symptoms listed in Tables 10.1–10.4 are reasonably specific.

Table 10.4 Diseases associated with nervous symptoms in adult cattle

General symptom	Specific symptoms	Suspect
Excitement/frenzy	bellowing, muscle spasms convulsions	hypomagnesaemia several poisons BSE
Disorientation/ delirium	head pressing circling blindness incoordination	listeriosis ketosis overeating disease lead poisoning vitamin A deficiency BSE
Depression but no other symptoms		subacute ketosis/fatty liver (early hypocalcaemia, botulism)
with fever or other symptoms	multiple	most systemic disease states

Others, such as inflammation of the udder, coughing, discharge from the eye, tend to be even more specific and do not merit discussion here. There remain the four most common symptoms of general malaise, fever, inappetence, dullness and loss of body condition. These may not be very helpful in establishing a diagnosis but the first two, at least, need to be given thought if we are to understand what it is to be a sick cow.

Fever

Fever is a controlled elevation of body temperature induced by chemical substances called *pyrogens* which can come directly from invading bacteria but which are more often derived from damaged tissues, or from the cells of the body's immune system (neutrophils, macrophages, etc.). These pyrogens reset the cow's thermostat in the hypothalamus. When we take rectal temperature with a thermometer we may classify fever, somewhat arbitrarily, as:

	°C	°F
rectal temperature, normal	38.2	101.5
moderate fever	38.2–39.5	101.5–103.1
severe fever	over 39.5	over 103.1

In the early stages of fever, the cow increases body temperature both by elevating heat production and reducing heat loss. Thus she shivers and reduces blood flow to the extremities so that her ears and feet may feel cold even though deep body temperature is elevated. When body temperature has stabilised her metabolic rate will remain elevated. Her extremities, especially her ears, may now feel hot but she may continue to shiver if it is necessary to maintain body temperature at the elevated level. The fevered cow is dull, off her food and may pass excessive amounts of urine. The combination of high metabolic rate, inappetance and high fluid loss can cause her to lose condition very rapidly.

Since fever is a natural manifestation of the body's defences to infection or tissue damage it is, presumably, not in itself a bad thing although I have yet to read a really convincing argument in its favour. Since most fevers in cattle are caused by infections, treatment is usually based on antibiotics which destroy or inhibit the original cause of the fever and so allow the natural processes of healing to take their course. We should, however, acknowledge that when we ourselves are fevered we are usually more concerned with treating the symptom, for example with aspirin, than attempting to remove the cause. If a cow has a fever so severe as to be an immediate threat to life she may be hosed down with cold water. The need for this is rare. What is far more common is the cow with a moderate, persistent fever who lies in a cold box shivering prodigiously in order to maintain a body temperature of, say, 39.5°C. Continuous shivering not only accelerates the loss of body tissue but can become exhausting. Such a cow will benefit from a rug, deep straw bedding or any other means of reducing heat loss, so making it easier for her to maintain an elevated body temperature. The use of drugs for symptomatic treatment of fever in dairy cows is probably not cost-effective since it will not accelerate the healing process. We can, however, without anthropomorphism, assume that the fevered cow is feeling rotten and nurse her accordingly.

Inappetence and anorexia

Inappetence refers to reduced food intake and anorexia means that food intake has stopped altogether. In Chapter 2 we considered what persuades a healthy cow to start and stop eating. Inappetence and anorexia in the sick cow are governed by signals from the same mechanisms that control hunger and appetite. A cow may refuse food if:

(1) It hurts to eat.
(2) The rumen or abomasum is already distended by undigested material.
(3) The chemicals circulating in the blood stream indicate satiety rather than hunger to the appetite centres in the hypothalamus.

Eating can be a painful process in diseases such as wooden tongue, lumpy jaw and traumatic reticulitis. Also, although less obviously, a cow with sore feet may prefer not to eat if it means walking to the feeding area where she has to stand and compete with other cows. We have been able to sustain food intake and thus milk yield in cows with severe foot lameness in early lactation by allowing them to lie, between milkings, in deep straw in a calving box and bringing the food to them.

An excess of undigested food can reduce food intake in cases of rumen or abomasal impaction, obstruction of the small intestine or spiral colon, or neuromuscular disorders leading to a reduction in the rate of propulsion of food through the gut.

When an animal is fevered or poisoned by toxins arising from bacteria or cell damage, the concentration of certain chemicals, particularly amino acids, in the blood rises in a way rather similar to that following a high protein meal. Moreover, stress has altered the balance of the endocrine system from anabolic processes, involving the uptake of VFA and amino acids into milk and body tissues under the control of somatotrophin and insulin, to catabolism, involving the release of glucose and amino acids into blood under control of the hormones of the adrenal cortex. All these factors signal to the hypothalamus that no more food is required until the chemicals already in circulation have been processed. The animal with liver damage may become completely anorexic because it cannot even process the products of its own tissue metabolism, let alone the products of digestion.

Prolonged inappetence presents the ruminant with two special problems. Firstly, the loss of fermentable carbohydrate from the rumen inevitably means a loss of ruminal microbes and thus a reduced ability to digest food when the cow elects, once more, to eat. Secondly, any dairy cow that is in milk or late pregnancy (i.e. nearly all of them) and goes off food is bound to develop some degree of ketosis. Ketones are, paradoxically, chemicals that inhibit appetite, especially for concentrate foods that are rapidly digested and absorbed. Once a cow goes off food therefore it is often difficult to rekindle her appetite. Most veterinary surgeons will dispense stimulant drenches which may work for reasons that I don't understand. Cyanocobalamin (vitamin B_{12}) works sometimes, partly for reasons that were discussed in Chapter 7. The stockman can also help to titillate the cow's appetite with fresh green food, succulents, or a warm mash which might contain sugar beet pulp, a cereal/bran mixture, molasses, and even stout!

Notifiable diseases

The UK Animal Health Act (1981) empowers the Ministry of Agriculture Fisheries and Food, through the State Veterinary Service and Local Veterinary Inspectors, to record and control the movement of livestock both within the country and overseas. The Act requires owners of cattle to identify animals by means of an ear tag, to report all cases of sudden death or abortion and to inform the Ministry or the Police if they suspect that any animal is suffering from a *notifiable disease*. These, at present, include anthrax, brucellosis, enzootic bovine leucosis, foot-and-mouth disease, tuberculosis and warble fly.

The Ministry has elected to make these diseases notifiable so that it can control them at a national level. Brucellosis and tuberculosis are controlled by an eradication policy because, in addition to the harm they do to cows, they can be transmitted to man in milk and other body fluids and cause severe, chronic disease or death. Some forms of salmonellosis are also transmissible to man and all diagnosed cases of salmonellosis in farm animals should also be reported according to the Zoonoses Order (1975). Foot-and-mouth disease and enzootic bovine leucosis are controlled by eradication policies because they are not established in the UK and it is deemed good policy to keep them out. Warble fly has been endemic but the natural history of the disease is such that it has already been reduced to a tiny fraction of its former incidence by a national control programme that only began in 1978. Salmonellosis is endemic and practically impossible to control on a national basis. Notification of the Ministry is, at present, intended mainly to identify possible sources of infection for control of the disease in man.

Brucellosis

Brucellosis in cattle is caused by infection with the bacterium *Brucella abortus*. As the name implies, the most serious feature of the disease is abortion at about the seventh to eighth month of pregnancy. The aborted calf and especially the placenta are highly infectious and the disease can be transmitted to other susceptible cows that investigate the aborted calf or come into contact with secretions from the infected cow. This can produce a catastrophic 'abortion storm' in a herd that lacks immunity. The cow that aborts is likely to have a chronic uterine infection, possibly a retained placenta, and may be very difficult to get back into calf. Most cows develop an effective immunity against further infection but some may abort on two or three occasions. Brucellosis used to be an occupational hazard of vets, dairy farmers and slaughtermen, especially in knacker yards. The disease in man is known as 'undulant fever', a seemingly endless succession of bouts of sickness having symptoms similar to influenza – fever, muscle and joint pains, dullness or sometimes severe depression. Between attacks the organism survives inside body cells where it is very difficult to treat with antibiotics. Humans can develop brucellosis through drinking infected milk but this is extremely rare as the organism is killed by pasteurisation. Most people are infected through contact with cows that have aborted.

Control of brucellosis in the UK began with a policy of vaccinating calves between three and six months of age. This reduced the incidence to the point where an eradication policy became possible. Breeding stock were blood tested and infected animals destroyed until, in time, vaccination could be discontinued. The national herd is now (almost) brucellosis free but, of course, has no immunity if the infection should reappear. The price of freedom from brucellosis is external vigilance, which is why the law requires farmers to report all cases of abortion in cattle to the State Veterinary Service so that samples of blood, milk, placenta or uterine discharge can be tested for brucellosis.

Tuberculosis

Bovine tuberculosis is caused by the organism *Mycobacterium bovis*. Cattle become infected by eating contaminated material or inhaling infected droplets. Initially the disease infects one or more lymph nodes near the point of entry, where it forms a slowly developing 'cold' abscess or tubercle. In many otherwise healthy cows (and people) the disease may not progress beyond this point but the animal is 'sensitised' to the antigens in the bacterium. However, before bovine tuberculosis was controlled, the infection would, in many cows, progress to the udder, enter the milk and so become a source of infection to man. The generalised or miliary form of tuberculosis, especially in children, was primarily due to drinking infected milk; tuberculosis of the lung, or 'consumption', was more often transmitted by droplet infection from human to human.

The eradication scheme for bovine tuberculosis has been a triumph of public health but it is not entirely without problems. It is based on the identification of cows which are sensitised to the antigen or 'tuberculin' in *Myco. bovis*, thereby indicating that they have been exposed to infection although they are, in all probability, not sick. There are however several other *Mycobacteria* such as that responsible for avian tuberculosis which do no harm to cattle or humans but which can produce local reactions in cattle sensitised to *Myco. bovis*. Tuberculin testing in the UK is therefore based on the 'double intradermal test': the hair is clipped from two sites on the skin of the neck; skin thickness at both sites is measured using calipers, and recorded; avian and bovine tuberculin are then injected *into* the thickness of the skin at top and bottom sites. Three days later the sites are remeasured with calipers to assess what swelling, if any, has occurred. If there is no significant swelling or 'nodule' at either site, the cow is not sensitised to either tuberculin. If the upper nodule is equal to, or greater than the lower, it is assumed that the cow is sensitised to a *Mycobacterium* other than *Myco. bovis*. If the lower nodule is substantially greater than the upper, the cow is deemed a reactor.

Despite these control measures it has proved impossible to eliminate real (or apparent) tuberculosis entirely from the UK. There are several reasons for this:

(1) Other species, e.g. badgers, farm or feral cats, can act as reservoirs of infection.
(2) In the early stages of the disease cattle may not react positively to the tuberculin test but can still infect others.
(3) No trace of tuberculosis can be found in some reactors which introduces the infuriating possibility of false positives.

This stubborn, last-ditch resistance of bovine tuberculosis to complete eradication has caused distress to vets, farmers and lovers of wild life who have witnessed the slaughter of cows, which are apparently healthy, and badgers, which are often very sick. There can be no justification for relaxing a policy that has done so much good. It is just possible however that we may, in time, be able to modify the existing policy to good effect.

Bovine spongiform encephalopathy (BSE)

Bovine spongiform encephalopathy, BSE or 'mad cow disease' was first identified in 1986. The early signs of apprehension, hyper-excitability and irrational behaviour progress to severe inco-ordination and mania. In many cases this is accompanied by a substantial fall in milk yield and body condition. The average age at onset of symptoms is 5 years (Wilesmith and others, 1992). Histolog-ical examination of the brains of affected animals reveals degeneration of nervous tissue with the formation of holes (hence 'spongiform') and characteristic fibrils or streaks of useless protein. The similarity of this picture to that of scrapie in sheep suggested early on that they may be the same condition. Scrapie in sheep has existed for at least 200 years. Epidemiological studies by Wilesmith, Ryan and Atkinson (1991) make it clear that the onset of the BSE epidemic can be linked to the change in the process for rendering meat and bone meal that occurred in 1981 (Chapter 6). The infectious agent in sheep with scrapie, as yet unidentified, was eaten by cattle (initially in the form of calf pellets) in sufficient amounts to provoke development of the disease while in the prime of life. Cattle will have consumed the scrapie agent before 1981 and it is probable that there were occasional undiagnosed cases before 1986. It is therefore almost certainly not a new disease but one which has assumed epidemic proportions because of a change in the intensity of challenge with the infective agent.

The evidence so far continues to support the conclusion that the only significant mode of infection is by eating infected material which implies that the number of recorded cases should start to fall rapidly in 1993, five years after the ban on inclusion of ruminant-derived protein in cattle diets and be close to zero within another five years. One cannot feel entirely confident about this, not least because it is known that scrapie can be transmitted vertically from mother to offspring and horizontally from ewe to ewe, probably by contact with infected placentae. At present hundreds of calves from infected and uninfected cows are being kept under close watch to discover whether there is any evidence for vertical transmission. So far so good.

The big worry concerning BSE is of course the possibility that if cattle can become infected by eating material from infected sheep we may become infected by eating material from infected cattle. This fear is stoked by the similarity between BSE and (rare) spongiform encephalopathies in man, Creutzfeldt-Jakob disease and especially Kuru, a condition that arose in Papua New Guinea due to the ceremonial practice of eating ancestors. Girls, who were the favoured descendants, were given the prized portion, i.e. the brain, and died first. The most reassuring evidence is that there is absolutely no suggestion of a link between Creutzfeldt-Jakob disease in man and scrapie in sheep even among those people who regard sheep brains as a delicacy. It would be foolish to assert that the risk to man from BST is nil, although no more foolish than some of the paranoid suggestions that we are gambling with the lives of an entire generation. Nothing in life is without risk but I can think of few safer pursuits than eating beef in moderation.

Enzootic bovine leucosis (EBL)

This disease has never become established in the UK. It is a virus condition that produces tumours in the lymph nodes eventually, since the usual incubation period is 4–5 years. The disease is endemic in countries such as the USA and Canada. Some Canadian Holstein cattle imported into the UK have responded positively to blood tests for the presence of the virus, although to my knowledge no commercial cow in the UK has ever shown clinical signs of the disease. It is relatively easy to establish and maintain EBL-free herds but the disease is a major restriction on the movement of Holstein cattle from North America to the UK.

Warble fly

The emergence of tens or even hundreds of maggots 25–30 mm long from the backs of perhaps half the cows in a herd during the months of March to July used to be one of the more revolting aspects of dairy farming, and the highly successful campaign to eradicate this infection has been welcomed by all parties. The maggots are the larvae of two species of warble fly, *Hypoderma bovis* and *Hypoderma lineatum* which lay their eggs on the hairs of the legs and abdomen of cattle during summer and early autumn (usually May to September). The larvae hatch after 4–5 days, burrow into the skin and migrate through the body of the cow reaching preferred sites in the oesophagus or around the spinal cord by December. They pause here until about February when they migrate to the tissue under the skin of the back, bore an air hole for themselves and grow into the full sized maggot that emerges in early summer. On leaving the cow, the larvae pupate and emerge 3–5 weeks later as adult flies.

The presence, especially the sound, of the warble fly in summer can disturb cattle to the point of panic when one beast or the whole group run madly about with their tails high in the air. This behaviour, known as 'gadding', has always caused me to wonder why it is that cattle who will tolerate massive infestations by the head fly with stoic indifference are so driven to panic by the sound of small numbers of flies *that do not hurt them at the time*. There is little doubt that in the late stages of migration and development the larvae do cause some pain and loss of performance. In very rare cases an errant larva can enter the spinal cord and produce paralysis. In all cases the larvae reduce the quality of the hide and carcass. The official control policy is to treat the cattle with a systemic insecticide between 15 September and 30 November, i.e. after the larvae have entered the body but before they begin to concentrate around the oesophagus and spinal cord. The insecticide is usually poured onto the back of the animals and absorbed through the skin. This means of administration has a superficial similarity to the old-fashioned process of scrubbing the back with derris solution in the spring when the larvae were emerging, but the site of application is immaterial – the insecticide enters the blood and all body secretions, destroys the larvae before they begin to do substantial damage and, incidentally, protects the animal against two other parasitic conditions of the winter – mange and lice.

Foot-and-mouth disease

Foot-and-mouth disease (F&M) is established in most areas of the world with the exception of North America, Australia, Eire and UK. The virus is transmitted by direct contact, by inhalation of infected aerosols, which can in special circumstances transmit the infection up to 100 km (e.g. across the English Channel), via carriers such as birds or insects, or in feedstuffs (e.g. pig swill). The virus affects the cloven-hoofed animals, cattle, sheep and pigs. In cattle it causes an initial acute phase of extreme sickness with almost complete loss of appetite and milk production. Ulcers appear in the mouth and feet just above and between the hooves. If permitted, most cattle recover but convalescence is prolonged and the equivalent of one lactation is lost. Given the fact that morbidity is close to 100% (i.e. nearly all animals exposed to infection do fall sick) it is only too clear that uncontrolled F&M would be a disaster for the National Dairy Herd.

It is however valid to ask whether the disease should be controlled by a slaughter policy or by vaccination. There are vaccines against the various strains of F&M and these are used with effect in countries where the complete eradication of the virus is impossible. The limitations to control by vaccination are:

(1) Some vaccinated animals can become symptomless carriers and so infect others.
(2) The immunity conferred by vaccination is short-lived and re-vaccination must be carried out at intervals as short as 4 months. This is extremely expensive.
(3) Adoption of a vaccination programme acknowledges that the disease is established within the country. This immediately destroys the export trade of live cattle to those countries that are F&M-free.

The UK policy for eradicating F&M is based on slaughter of infected herds plus restrictions on the movement of all livestock within a 10-mile radius. This is Draconian but usually very effective at restricting the spread of disease. However, the 1967–68 outbreak which started in the north-west of England almost got out of hand, largely because of the impossibility of preventing aerosols containing F&M virus being carried from farm to farm on the wind. If an outbreak of similar magnitude occurs in future the slaughter policy may have to be reinforced, temporarily, by vaccination.

Salmonellosis

The two species of salmonella most often responsible for disease in cattle are *S.typhimurium* and *S.dublin*. The former can infect most mammals including man, the latter is almost entirely specific to cattle. The organism is practically impossible to eliminate from infected farms since it can survive for long periods both outside and inside animals without causing disease unless their resistance is upset by a stress such as calving, transport or dietary disturbance. The problem of salmonellosis, especially that due to *S.typhimurium*, is greatest in young calves who may be infected by their mothers or in transit from their farm of origin to rearing unit. *S.dublin* can cause outbreaks of diarrhoea in well grown heifers and adult cows and is a relatively common cause of abortion (Chapter 12).

Other infectious diseases

The next two sections deal very briefly with some infectious and parasitic diseases that can affect cows and heifers. There are no detailed descriptions of the clinical signs and pathology of these diseases. In each case I concentrate only on features that affect the transmission of these diseases and therefore their control on the farm.

Bovine virus diarrhoea (BVD)

The virus responsible for BVD (or mucosal disease) has mild and virulent strains which can affect cattle of all ages and cause disease that ranges in severity from mild inappetence and diarrhoea to massive erosion of the gut wall leading to profuse diarrhoea, rapid weight loss and early death. Typically, a herd of cattle may be infected for the first time when a carrier is purchased into the herd, especially during the winter months when cows are housed. On this first occasion most of the cows may experience diarrhoea, inappetence and a sharp drop in milk yield for a few days and then recover. Some may abort. Very rarely an individual may develop severe, uncontrollable diarrhoea and die. Most of those that recover become immune but carriers persist, the disease becomes endemic in the herd and continues to affect heifers and purchased cattle.

If this were all it would be bad enough. There appears however to be a further complication that if cows are infected within the first six months of pregnancy with the mild strain of the virus it enters the foetus before its immune system is mature enough to recognise the virus as 'not-self' and manufacture antibodies against it. This weak strain of the virus usually does little harm. However, if calves are exposed after birth to the virulent strain their immune system is unable to recognise it as a threat, they are unable to mount a defence and the disease is rapidly fatal. Thus calves which are born to cows infected with BVD in the first six months of pregnancy and which stay in or move to a farm where BVD is endemic are unlikely to survive long enough to enter the milking herd or be sold as meat animals.

One way to avoid the worst problems of BVD in a herd where the disease is endemic is to ensure that all heifers are exposed to infection before becoming pregnant for the first time. This is easier said than done. It is also possible to eradicate the disease based on tests to discover which animals have antibodies and which have virus but nò antibody. Details of control on individual affected farms are beyond the scope of this book and a matter for discussion between the farmer and his veterinary surgeon.

Johne's disease

The main clinical feature of Johne's disease is a profuse, persistent watery diarrhoea. Initially the cow is bright and has a good appetite but the massive loss of water and electrolytes causes a rapid loss of body condition and progressive weakness. The disease is caused by *Mycobacterium johneii* which, like the bacteria responsible for tuberculosis and brucellosis, can become established within the cells of the body and so become very resistant to antibiotics. Affected cows should be culled as quickly as possible. Although it is only adult cows that show the disease they are, in fact, resistant to infection which can only become established in calves under 6 months of age. It lies dormant until provoked in later years by a stress such as calving. The organisms cause a progressive thickening of the lower small intestine and colon which progressively reduces the cow's capacity to absorb water, electrolytes and nutrients from the gut.

The disease has become much less common in dairy cows than formerly since it was recognised that it could be controlled by keeping young calves away from the faeces of clinically affected cows. Young calves can be vaccinated against *M.johneii* in high-risk herds but this requires a special licence from the Ministry of Agriculture since vaccinated animals react to the tuberculin test based on the closely related bacterium *M.bovis*

Clostridial diseases

The two main clostridial diseases of cattle are tetanus and blackleg. Both conditions are sporadic but likely to be fatal since animals die before treatment can be put into effect. Two other conditions, black disease and botulism, are rare. They are all caused by the family of bacteria called *Clostridia* which form spores and so survive in the soil almost indefinitely. The infections cannot therefore be eradicated but can be controlled most effectively by vaccination. Vaccination against clostridial diseases is considered to be essential preventive medicine by sheep farmers who may, however, not bother to vaccinate their much more expensive cattle grazing the same pastures. Sheep are more prone to clostridial diseases than cattle, partly because their grazing habits (and the grazing pressure imposed on them) bring them into much closer contact with the soil and the clostridial spores therein. The decision as to whether or not it is cost-effective to vaccinate cattle tetanus and blackleg is always a calculated risk and one that should be based on local knowledge of the incidence of the disease.

Parasitic diseases

The internal parasites of cattle can, for practical purposes, be divided into those like liver-fluke which require a second, intermediate host to complete their life cycle, and those like the lungworm *Dictyocaulus viviparous* which survive and develop in the pasture and re-infect cattle directly. In the former case it is possible to eliminate the disease by removing the intermediate host. In the latter case it may be assumed that the parasite cannot be eliminated except by keeping cattle off the pasture for an unrealistic period of at least one year. Control is therefore a matter of coming to terms with the parasite. It is, in fact, the essence of a successful host–parasite relationship that each comes to terms with the other. The roundworms that reside in the gut and lungs of cattle are most likely to do harm in the calves' first summer at pasture. Thereafter the calves develop a degree of immunity, and parasitism is unlikely to be a problem in the adult cow unless her resistance is impaired by malnutrition. Infestation with gastrointestinal roundworms (helminths) can therefore cause diarrhoea, malabsorption, dehydration and loss of body condition in calves during their first summer and occasionally in malnourished beef cows, but is unlikely to affect dairy cows to a significant extent. It was thought for a while that even a few helminths in adult cattle must be doing *some* harm and that healthy, lactating cows would therefore perform even better if treated regularly with anthelmintic drugs. There may be a small effect on production in some cases but not enough to justify the cost of the drugs and the labour involved.

Lungworm

The parasite *Dictyocaulus viviparous* can cause severe pneumonia ('husk' or 'hoose') in young calves at pasture. The mature lungworms live in the bronchi, interfere with the normal movement of air during respiration, and cause moderate to severe damage to the lungs. In a few cases, affected animals develop a high temperature and severe respiratory distress and may die within a few days. In the more common, chronic form of the disease, calves cough repeatedly and lose condition rapidly. The disease is extremely widespread in the UK and some form of prevention is essential for calves in their first summer.

Eggs from the mature lungworms are coughed up and swallowed. In the gut they mature into larvae which are shed in the faeces and mature further on the grass to the point where they are able to re-infect cattle. They can survive the winter and infect young cattle in the following spring. The best form of prevention is to immunise calves before turnout with two doses of a vaccine such as Dictol, which is made by irradiating larvae to the point where they are unable to mature into adults but still induce immunity. Natural exposure to lungworms later in first and subsequent seasons at pasture is sufficient to maintain immunity, so vaccination does not need to be repeated. Some farmers, however, find it cheaper to give one or more preventive doses of an anti-parasite drug such as Invermectin which is usually cheaper and has the additional merit of killing most other internal and external parasites.

Liver-fluke

The liver-fluke is a small flat worm that spends its adult life in the liver and bile ducts where it sucks blood. Severely infested animals become anaemic and emaciated. Eggs are shed into the faeces, hatch in the warm, wet weather of early summer and penetrate the snail *Lymnea truncatula*, whose natural habitat is poorly-drained, chronically muddy areas around streams and water holes. The snails do not thrive in highly acid peat bogs but can multiply if the land is limed to improve its potential as a cattle pasture. The intermediate stages of the parasite mature and multiply inside the snail, are cast onto the pasture and ready to re-infect cattle by about September. It then takes about three months for the mature flukes to develop so that the clinical disease is usually a problem of mid-winter.

To control the liver-fluke, wet areas of pasture should be drained or fenced off. If this is not possible and if the summer has been wet enough to encourage multiplication of both snails and the intermediate stages of the fluke, cattle can be given a preventive dose of a fluke-killing drug in December when the flukes are inside the cow but before they have begun to do any damage.

Tick-borne diseases

Redwater is a disease of cattle caused by a single-cell parasite *Babesia divergens* with requires as an intermediate host the cattle tick and so can only affect cattle in those areas inhabited by ticks. In the UK these are mostly the highlands and moorlands. The parasite multiplies in the blood stream and destroys red blood cells, releasing haemoglobin into the urine which, during acute phases of the disease, is said to have the appearance of port wine. The anaemia that develops causes severe distress, the heart and lungs labouring to sustain the flow of oxygen to the tissues. Some animals die within a few days and the rest take several weeks to recover. In parts of the world where the tick is endemic, vaccines against *Babesia* can be used but these are not available in the UK.

The only other tick-borne disease of importance in the UK is tick-borne fever, caused by a rickettsia which attacks white blood cells. It is not usually a severe disease but it can reduce the immunity of cattle to concurrent infection.

Prevention and treatment of infectious disease

Prevention

Various methods for the prevention of infectious diseases have been outlined in this and previous chapters. These resolve into four basic strategies, the method of choice being determined by the nature of the disease.

(1) Eradication – complete elimination of the causative organism: (a) from the country, e.g. foot-and-mouth disease, (b) from the farm, e.g. tuberculosis.
(2) Vaccination – specific protection against organisms that cannot be eliminated from the farm, e.g. clostridial diseases, husk, some viral infections (infectious bovine rhinotracheitis in calves).
(3) Hygiene and management – control of endemic diseases by management of the environment so that: (a) the challenge to the animal is as low as possible, (b) the resistance of the animal is as high as possible.
(4) Preventive medication – the use of long-acting drugs to destroy or inhibit bacteria or parasites *before* they cause clinical disease, e.g. 'dry-cow therapy', warble fly control.

Treatment

The natural defence mechanisms of a well-nourished animal are able, in time, to overcome most infections. The natural history of (almost all) infectious disease proceeds therefore towards a cure. Treatment of infectious disease involves:
(1) Drugs which destroy or inhibit organisms which are the direct cause of disease (primary pathogens) or which can exploit the tissue damage created by primary pathogens and make matters much worse (secondary, or 'opportunist' pathogens).
(2) Supportive therapy and nursing:
   (a) drugs to reduce the more distressing symptoms of infection,

(b) nutrients, electrolytes, etc., to assist the animal to maintain homeostasis and to accelerate healing,

(c) thermal and physical comfort.

Some general points need to be made concerning the use of antibiotics and related drugs for specific anti-microbial therapy. Antibiotics such as penicillin and anti-microbial drugs such as the sulphonamides and nitrofurans interfere with the metabolism of certain specific bacteria in ways that kill them or inhibit their growth and multiplication. Each antibiotic is effective against a specific, limited number of bacterial species and strains. When a vet chooses an antibiotic to treat a particular case he makes his decision on his knowledge, or expectation, of the species and strain of bacterium responsible for that disease. Before embarking on a course of treatment for, say, mastitis or salmonellosis, he may take a sample of milk or faeces and run sensitivity tests to determine which antibiotics are most effective at inhibiting bacterial growth. However, he will, almost certainly, have to start treatment before the result of the sensitivity test is known and therefore cannot guarantee to get it right first time. I hear all too commonly farmers (or the general public) say, 'My vet (doctor) gave me this antibiotic but it didn't work so now he's trying a stronger one.' This is a nonsense. Strength has nothing to do with it. An antibiotic simply is, or is not, appropriate.

Table 10.5, condensed from Blood and Radostits (1989), lists the main groups of antibiotics and other microbial drugs and their efficacy against some of the common bacterial infections of cattle. Antibiotics are *not* effective against viruses. When antibiotics are given to a cow or calf suffering from a primary virus infect such as pneumonia caused by respiratory syncytial virus, it is to prevent or control secondary damage caused by opportunist bacteria such as *Pasteurella haemolytica*.

The original form of penicillin was effective against *Streptococci*, *Staphylococci*, *Clostridia*, *Corynebacteria* and *Listeria*. In time some penicillin-resistant strains of *Streptococci* have developed and many penicillin-resistant strains of *Staph. aureus*. New, synthetic pencillins such as Cloxacillin have been developed to counter *Staph. aureus*. Other synthetic, penicillin-like drugs such as Amoxycillin and Ampicillin are also effective against *Pasteurella* and certain strains of *E.coli* and *Salmonella*. These are termed 'broad-spectrum antibiotics'. Streptomycin used to be effective against *E.coli* and *Salmonella*. However, these organisms present special problems in human and veterinary

Table 10.5 Sensitivity of some microbes to anti-microbial drugs

	Penicillin-G	Cloxacillin	Amoxycillin	Streptomycin	Tetracyclines	Chloramphenicol	Sulphonamide	plus trimethoprim	Nitrofurans
Strep.agalactiae	S	S	S		S	S	GS	GS	S
Other streptococci	GS	S	S		GS	S	GS	GS	S
Staph. aureus	OR	GS	OR	OR	OR	GS	OR	GS	GS
Clostridium spp.	S	S	S		S	S	GS	GS	
Corynebacterium spp.	S	S	S	S	S	S			
Listeria	S	S	S		S	S	GS	GS	
E.coli,gut			GS	GR	GR	OR	OR	GS	GS
E.coli,milk			GS	OR	GS	GS	OR	GS	GS
salmonella spp.			GS	OR	OR	GS	OR	GS	GS
Pasteurella spp.	OR	OR	S	GR	OR	S	OR	GS	GS
Actinobacillus spp.	S		S	GS	GS	GS	GS		
Brucella abortus			GS	S	S	S		GS	GS
Leptospira spp.	S			S	S	S			
Campylobacter spp.				S					
Mycoplasma spp.					S				
Actinomyces spp.	S		S	S				GS	GS

S = sensitive, GS = generally sensitive, OR = often resistant, GR = generally resistant. Blank spaces indicate that the organism is resistant (or information is unavailable). Condensed from Blood, Henderson and Radostits (1983).

medicine because each species has hundreds of different strains, many of which are resistant to a wide range of antibiotics. Moreover, antibiotic resistance can be transferred between strains and even between the two species. Repeated use of antibiotic therapy for these infections kills off the sensitive strains of bacteria but, in doing so, permits the multiplication of antibiotic-resistant strains. Thus streptomycin is now practically useless as a treatment for colibacillosis or salmonellosis.

Tetracyclines remain excellent broad-spectrum antibiotics and are used as first-choice for a wide range of infections. However, many strains of *Staph. aureus* and *Salmonella* are now resistant. Chloramphenicol has an extremely broad spectrum of activity but must be used with extreme discretion in veterinary medicine to avoid the development of chloramphenicol-resistant strains of *Staph. aureus* and, especially, *Salmonella* species which can infect and (rarely) kill man. Other anti-microbial drugs which are not antibiotics include the sulphonamides, especially in association with trimethoprim and the nitrofurans. These all have a broad spectrum of action although not so broad as that of chloramphenicol.

Antibiotics are usually administered to adult dairy cows by injection intravenously or intramuscularly, or direct into the udder via an intramammary tube. Intravenous injection dispenses antibiotic rapidly throughout the body. Intramuscular injections usually achieve a more gradual build-up but longer duration of the systemic effect. When administering antibiotics it is, of course, essential to withdraw milk from the bulk tank for the prescribed period. It is also essential to complete the prescribed course even if the cow appears to have already made a complete recovery so as to kill *all* the pathogenic organisms (e.g. *Staph. aureus* in the mammary gland) and reduce the probability of a relapse.

Finally, we must remember that since antibiotics kill bacteria they will affect the normal microbial population of the rumen. Some loss of appetite is to be expected after injection of a systemic antibiotic to a cow. Very occasionally, cattle food is contaminated by antibiotics intended for inclusion in pig or poultry rations. The effects can, at worst, be catastrophic, leading to an almost complete loss of microbial activity in the rumen, anorexia, a precipitate fall in milk yield and, sometimes, death.

Part IV
Breeding and Fertility

11 Breeding

Every dairy farmer wants the best cows for the job. He is encouraged to breed his replacement heifers using semen from bulls of superior genetic merit on his most productive cows and so ensure that the average performance of his herd improves with each successive generation.

The performance of any animal, be it the capacity of a dairy cow to produce milk solids or the speed of a racehorse over the $1\frac{1}{2}$ miles of the Derby course, is the consequence of all the effects of the environment (E) in which it has lived since conception on the genes that it has acquired from both its parents. The actual appearance of an animal in a qualitative trait such as colour, or its performance in a quantitative trait such as milk yield, is the *phenotype* (P). The extent to which each trait is determined by the genes an animal inherits from its parents is the *genotype* (G). The phenotypic value of any trait is the sum of the effects of the genotype and the effects of the environment. Thus:

$$P = G + E$$

For a qualitative trait such as colour or the presence or absence of horns it should be obvious that phenotype and genotype are usually synonymous. Conformation (or 'type') is largely determined by genotype but influenced to some degree by environment. The phenotypic expression of important quantitative traits such as milk yield and composition is greatly influenced by the environment. For example, the daughter of the highest yielding Holstein cow and the best Holstein bull in the UK will not produce very much milk if shipped to central Africa and fed on a diet consisting of hay. The environment prevents her from realising her genetic potential. Indeed, for a variety of reasons, the 'superior' cow might, on this ration and in this heat, produce less milk than an African cow of humbler parentage. This simple example illustrates the principle of genotype/environment interaction. The highly selected Holstein cow is not necessarily the superior animal, merely one that is more appropriate to a husbandry system where high quality food is available and economic forces (especially fixed costs) favour the cow that produces the most milk (see Table 1.9).

The steady and continuing rise in *average* lactation yields of milk solids per cow in the intensive dairy farming areas of the developed world has been achieved by parallel improvements in genotype and nutrition. Any improvement in nutrition, e.g. substitution of good silage for hay, favours the cows that have the genetic potential to produce more milk. Selection of these cows for breeding replacement heifers improves average yields for the whole herd which then shows a further production response to increased concentrate feeding, which, once again, reveals the most productive cows – and so on.

Every dairy farmer should therefore have a breeding policy designed to achieve steady improvement in those traits associated with the quantity and quality of his product, the health of his cows and the profitability of his enterprise. Let us begin by considering what traits it might be worth attempting to improve by breeding, ignoring for the moment how easy or difficult each may be to achieve:

(1) Total yield of milk solids; especially butterfat and protein.
(2) Composition of milk, or percentage of butterfat and protein.
(3) Conformation of the cow:
    (a) to reduce the incidence of mastitis, lameness, trampled teats, etc., and so improve productivity and longevity;
    (b) to improve the value of her male calves as beef animals and, ultimately, her own value as a beef animal.
(4) Reproductive efficiency. In other species such as sheep this may involve selecting for twinning. In cattle it would normally involve selecting cows and bulls for ease of calving and rapid return to oestrus.
(5) Food conversion efficiency – increased milk output per tonne of purchased concentrate food, or per MJ of utilised metabolizable energy.
(6) Disease resistance – reduced incidence of mastitis, lameness, etc.

The relative importance of each of these objectives varies somewhat according to the economic circumstances of the moment. Before quotas, the total yield of milk solids was undoubtedly the most desirable trait since it increased income relative to fixed costs for labour and buildings (see Table 1.9). Even now, when output is fixed, the farmer with the higher-yielding cows is likely to make more profit which will enable him to purchase more and more quota from his poorer (or wearier) neighbours. However, it is unreasonable to expect total milk sales within the developed world to increase substantially within the foreseeable future. Thus it makes more long-term sense to emphasise selection for milk composition and, especially, efficiency of food conversion.

How successful has been the contribution of breeding to the improved performance of dairy cows? At first sight, not very much. Table 11.1 presents records from the Milk Marketing Board for milk production in recorded herds in the ten years from 1975 to 1985 when the impact of quotas first began to be felt. The average improvement in milk yield was 1.5% and 1.2% per annum in British Friesian and British Holstein cows respectively (MMB, 1986) when the majority of heifers were sired by superior bulls as assessed by progeny tests. This rate of phenotypic improvement was (a) small and (b) essentially no different from that in Dairy Shorthorns and Jerseys where the range of progeny-tested bulls has been far less than that available to Friesian and Holstein breeders. This implies either that the potential for genetic improvement is limited or we have not been doing it very well. It should, in theory, be possible to improve the performance of dairy cows at a slow but constant rate for the foreseeable future, this rate being at least twice as great as that we have achieved over the last twenty years. I shall attempt to explain in simple terms some of the factors that govern and constrain the rate of genetic progress.

Table 11.1 Lactation trends by breed (Milk Marketing Board 1985/86)

	1975	1985	% change per annum
Milk yield (kg)			
British Friesian	4817	5593	1.5
British Holstein	5558	6562	1.2
Dairy Shorthorn	4213	5019	1.7
Jersey	3415	3849	1.2
Milk fat and protein (kg)			
British Friesian	333	392	1.6
Dairy Shorthorn	291	350	1.8
Jersey	298	345	1.5

(1975 figures for milk fat + protein in Holsteins are not available.)

Selection in theory

All animals born as a result of the fertilisation of the ovum of a female by the sperm of a male are genetically different, with the single exception of identical twins formed by splitting of a single fertilised egg. Genetic selection is merely the manipulation of these differences. Before domestication, natural selection within a species, together with random mutations, operated to ensure the survival and development of species, combinations of species and individuals within a species that were best fitted to a particular environmental niche. After a prolonged period of natural selection the phenotypic variation between individuals within a species in a particular environment becomes very small for those traits that are essential to survival like reproduction and food conversion efficiency. Thus all the wild rabbits in a particular wood will tend to reproduce and grow at similar rates. Their coats will also be of similar colour since the wild rabbit needs camouflage and so coat colour is critical for survival. Similarly, phenotypic variation within species in the visual appearance of wild birds is very small, although here visual recognition tends to be more important than disguise. In general, natural selection within a species tends to reduce phenotypic variation for traits important to survival and fitness for the environment. Variation between animals that permits them to fit satisfactorily into different environmental niches has been achieved largely by evolution of different species, i.e. animals so genetically distinct that they cannot breed with one another. The corollary to the tendency to reduce variation within a species for those traits that are important to survival is, of course, the premise that where considerable variation persists it is because the trait is of little importance to survival of the species.

Domestication of farm and pet animals largely freed them from the problems of seeking sufficient food, competing for a sexual partner and avoiding the ravages of predators. At the same time it allowed man to engage his fancy for manipulating the phenotype of his animals by controlling breeding. There is, in nature, no phenotypic variation within a species to compare with that between, say, the Great Dane and Yorkshire Terrier breeds of dog, and little to compare with that between a Hereford and a Holstein cow. The Hereford and Holstein have been developed from a common ancestor by selection over many generations to produce, in the former case, a medium-sized animal for the production of meat and fat from grass, but which produces hardly enough milk for one calf, and in the latter, a large cow that produces an enormous amount of milk but has little value as a beef animal. Cattle with these major, consistent phenotypic differences come, in time, to be recognised as distinct breeds and are lovingly preserved as such by their respective breed societies.

Most articles on livestock improvement draw a clear distinction between (a) crossbreeding and (b) selection within a breed. Cross-breeding implies mating animals from two phenotypically different selected lines to produce a hybrid that suits a particular system of production. The Hereford bull is crossed with the Friesian cow to produce a calf that can be finished for beef in about 18 months almost entirely off fresh and conserved grass. In this case, crossbreeding is practised because the calf is intended for a very different role from that of its mother. The same logic applies when a Suffolk ram is mated with a Welsh Mountain ewe.

Selection within a breed may be for a Hereford bull that grows faster or a Friesian cow that gives more milk. Improvement by selection is obviously slower than the benefit that can be obtained from cross-breeding *where it is appropriate*, since by crossbreeding one exploits genetic differences already achieved by selection over many genera-tions. There is very little, however, that can be achieved by crossbreed-ing within the dairy industry, mainly because mothers and daughters are expected to perform the same function in the same environment and dairy bulls are only really respected for their feminine traits, i.e. the ability of their daughters to produce milk. One could, in theory, adapt to quotas by crossing Friesians with Jerseys to increase the concentra-tion of milk solids, although it would obviously be easier to swap 10 Jersey cows for 10 Friesians.

All conventional forms of livestock improvement are achieved, therefore, by selecting for and exploiting differences between animals based on the natural variation that exists within the population. The differing traits of importance, such as milk yield, are all quantitative (i.e. differing in degree) as distinct from qualitative traits such as the presence of horns, where the distinction is absolute. The physiological capacity of a cow to give milk depends on a large number of factors such as number of alveoli in the udder, appetite, endocrine status, etc., and so is under the control of a large number of genes. It may be possible, in time, to manipulate certain aspects of milk production by genetic engineering (Chapter 13) if they are determined by one or a small number of genes. For the immediate future, selection will continue to be based on the amount of genetic variation that exists within the population and the mathematical probability that animals deemed to be superior for a particular trait will transmit some of that superiority to their progeny.

Variation and heritability

Let us consider first a simple, measurable trait that a bull might reasonably be expected to transmit to its sons, say 400-day weight. Within a population of bulls raised in a particular environment on a

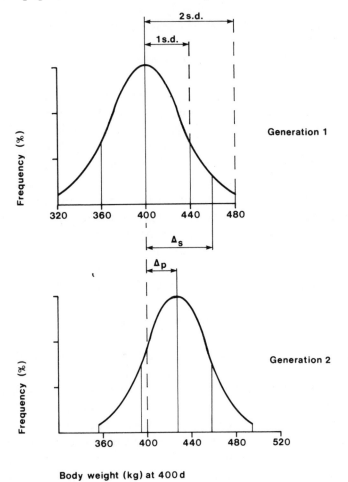

Fig. 11.1 Variation in body weight of cattle at 400 days, and the effect of selection when the sires from generation 1 were 60 kg above the mean (Δ_s), heritability is 0.42 and the progeny in generation 2 are, on average, 25 kg above the mean of generation 1 (Δ_p).

particular plane of nutrition, mean or average 400-day weight is 400 kg (Fig. 11.1). Individual weights at 400 days are distributed symmetrically about this mean value such that two-thirds of all bulls weigh between 360 kg and 440 kg, i.e. are within one standard deviation (s.d.) of the mean. 95% of bulls are within 320–480 kg (or 2 s.d.).

Suppose now that we select for breeding only those animals above 450 kg at 400 days on the basis of a performance test and that their mean 400-day weight is 460 kg (1.5 s.d. above the mean). We then *progeny test* their offspring in the same environment. Their mean 400-day weight is 425 kg and 95% weigh between 355 kg and 495 kg (Fig. 11.1). Mean weight has increased by 25 kg but the variation between the top and bottom 50% has decreased from 160 kg to 140 kg.

The *heritability* (h^2) of this trait is given by:

$$h^2 = \frac{\text{superiority of progeny mean over original herd mean}}{\text{superiority of parents over original herd mean}}$$

In this case

$$h^2 = \frac{425-400}{460-400} = 0.42$$

Approximate values for the heritability of various performance traits are given in Table 11.2. Yield of total milk, milk solids and body conformation are all moderately heritable ($h^2 = 0.25$), which means that these traits can be improved by continued selection. Traits such as longevity and calving interval (which describe the reproductive

Table 11.2 Estimated heritabilities of traits relevant to the performance of dairy cattle

Trait	Heritability
Milk yield (305 d)	0.25
Fat yield	0.25
Total solids yield	0.25
Fat (%)	0.50
Protein (%)	0.50
Body size	0.40
Calving interval	0.05
Longevity	0.05
Conformation	0.25

efficiency of the cow) have very low heritabilities. Maternal traits in all farm animals (such as number of piglets reared per year by sows) have notoriously low heritabilities which implies that natural selection and domestication have already achieved nearly all that can be achieved by genetic selection, which further implies that such phenotypic variation that remains is almost entirely of environmental origin. In practice, cattle and sheep maternal traits are usually enhanced by crossbreeding between two genetically different inbred lines in order to induce heterosis or 'hybrid vigour'. This is not really feasible in dairy cattle although the Hereford × Friesian hybrid makes an excellent suckler cow for beef production.

Figure 11.1, which is included mainly to illustrate the concept of heritability, also indicates a reduction in genetic variation as a result of selection in generation 1. In the theoretical case of a population of infinite size with no inbreeding, selection for a trait like milk yield which is governed by many genes causes some reduction in genetic variance in the first generation. Thereafter variation declines at a decreasing rate through subsequent generations until an equilibrium is reached, i.e. some variation persists for ever, the animals never become identical for the selected trait. It follows therefore that the response to selection is greatest initially and declines but, in a population of infinite size, eventually settles down at a final, asymptotic rate, theoretically for ever (Wray and Hill, 1989).

Until recently, the major constraint upon genetic improvement in dairy cattle has been the fact that superiority could only be distributed widely by the male (using artificial insemination) and his superiority could not be assessed until records of performance had been obtained from a sufficient number of his daughters. It is now possible to increase the contribution of the cow to genetic improvement using the technique of multiple ovulation and embryo transfer (MOET). With this technique each female may contribute (say) 16 heifer calves for test either after selection on the basis of her first lactation (adult MOET) or on the basis of her pedigree when she first begins to ovulate (juvenile MOET).

Theoretical rates of improvement in a trait such as milk production which has a heritability of 0.25 are given in Table 11.3.

Actual rates of genetic improvement in the UK Friesian herd have been about 0.3% per annum with the best herds approaching a rate of 1.6% p.a. In North America, progress has been more rapid with most recorded Holstein herds achieving 1.2 to 1.4% per year. Thus, at best, the rate of genetic improvement through conventional progeny testing

Table 11.3 Theoretical rates of genetic improvement in milk yield of dairy cows using different breeding techniques (from Hill, Thompson and Woolliams, 1990)

System	Rate of progress (%/year)
Conventional progeny testing	1.5–2
'Ideal' progeny testing	2.6–3.3
Adult MOET	2.4–3.1
Juvenile MOET	3.4–4.5

has been at a rate consistent with theory. Theory also argues that this rate of improvement could be doubled by the widespread application of MOET, especially using the juvenile scheme whereby heifers are selected upon their pedigree thereby reducing the generation interval to two years or less (Nicholas and Smith, 1983). These authors also suggested that cloning identical copies of superior animals by the technique of embryo splitting could achieve even greater gains.

This early optimism has given way to a more cautious view. Woolliams and Wilmut (1989) have argued that, even in the best of circumstances, cloning would not be likely to achieve significant gains in practice mainly because it would introduce problems of inbreeding. Radical reduction in genetic variation (e.g. within a single herd) could be a disaster if that set of near-identical animals proved to be particularly susceptible to an infectious or metabolic disease.

MOET schemes, without embryo splitting, are also more likely than progeny testing schemes to promote inbreeding when applied in practice. Woolliams and Wilmut (1989) argue that the most effective approach to MOET in practice is the adult scheme, and this is unlikely to achieve a more rapid rate of progress than the best of progeny testing schemes (Table 11.3). This raises serious questions as to the cost-benefit of the procedures involved in MOET (see Chapter 12) whether considered in terms of cow welfare or cash benefit for the farmer.

Correlated responses

The response to selection for milk yield is achieved by exploiting variation in a wide range of associated traits such as body size, food intake, mammary blood flow, etc. It should be obvious therefore that when selecting for a single but complex trait such as milk yield one must inevitably change other traits to a degree that is more or less closely correlated with the selected trait. For example, when selecting for increased milk yield one would expect to observe correlated increases in body size and food intake, and *may* record correlated increases in the incidence of mastitis and lameness. The statistical analysis of these correlated responses need not concern us here.

Table 11.4 (from the Milk Marketing Board) predicts some correlated responses of daughters of bulls selected for a range of simple production traits.

Selection for increased milk yield in this example achieves (over 8 generations) an increase of 410 kg in total lactation yield, 11.7 kg and 11.0 kg respectively in fat and protein yields, but small declines in fat and protein concentration (0.07% and 0.06% respectively). In other words, more milk has been produced but it has become a little wetter.

Selection, on the other hand, for increased *concentration* of fat or protein is very effective at achieving its primary objective (as expected since heritability is high) but causes a substantial reduction in total milk yield and, in three out of four cases, a reduction in output of milk solids.

Given the present price structure in the UK, the most useful trait for selection is milk fat *plus* protein since this causes maximum increase in

Table 11.4 Predicted correlated responses in daughters of bulls selected for a single trait

| | Correlated response | | | | |
| | Yield (kg) | | | Composition (%) | |
Selected trait	Milk	Fat	Protein	Fat	Protein
Yield (kg)					
milk	+410	+11.7	+11.0	−0.07	−0.06
fat	+329	+14.5	+10.4	+0.05	−0.02
protein	+363	+12.1	+12.4	−0.03	−0.02
Fat and protein	+360	+14.0	+11.8	+0.01	−0.02
Composition (%)					
fat	−151	+3.5	−1.8	+0.19	+0.07
protein	−207	−2.6	−1.6	+0.11	+0.12

yield of these two valuable ingredients without substantially affecting milk composition. It does appear that selection for increased yield of milk solids is associated with a small decline in protein concentration. Conversely, selection for increased protein concentration reduces yields of whole milk and all solids.

Just as the phenotypic variation between individuals in a single trait is a function of genotype and environment, so too is the observed phenotypic correlation between two traits. For example, there could be a phenotypic correlation between increased milk yield and environmental mastitis that could, on further analysis, be ascribed to 10 herds that switch to Holsteins without increasing the size of their cubicles. It is quite possible, in theory, to distinguish between genetic and phenotypic correlation but not always so easy in practice. For further details consult Falconer (1960) or the new edition of *Dalton's Practical Animal Breeding* by Willis (1991).

Methods of selection

The breeder seeking to improve the genetic merit of his cows makes decisions as to which cow should be mated with which bull on the basis of information as to the *relative* performance of his families of cows, their conformation or 'type', ease of calving, etc. within his own herd, and records of the *relative* performance of the progeny of bulls whose semen is available to him through the services of the Milk Marketing Board (MMB) or an independent cattle breeding company. Before going on to describe how this is achieved in practice in the UK, it is necessary to state, in simple terms, three more fundamental principles which are governed by the strict mathematical rules of population genetics:

(1) *The more traits chosen for inclusion in a selection programme, the slower is the rate of improvement for any single trait.* If, for example, a farmer wished to improve milk yield, composition, udder conformation, beef type and ease of calving all at once, and gave equal prominence to each trait, then he would be unlikely to achieve significant progress in any of these individual traits within his lifetime.

(2) *Selection for a single trait inevitably produces changes in other traits which may, or may not, be desirable.* Selection of cows only for their production of milk solids may, for example, produce animals with

progressively poorer conformation and an increased predisposition to lameness or mastitis.

(3) *An animals gets half its genes from each parent, one-quarter from each grandparent and one-eighth from each great-grandparent.* This is a profound statement of the obvious but is often overlooked in practice by animal breeders, particularly those who believe in selling animals on the basis of their pedigree. A famous grandmother would have to have demonstrated a superiority, relative to her own contemporaries, more than four times greater than that of her granddaughter in order to merit greater breeding value for the next generation of great-granddaughters. This is extremely unlikely. Pedigree records going back more than two generations may be historically interesting but are practically worthless for the purpose of selection.

Bearing these points in mind, there are three approaches which may be realistically adopted to the selection of bulls and cows for profitable, healthy milk production:

(1) *Tandem selection.* The breeder selects for a single character, say yield of milk solids, until it reaches an acceptable level, then switches to a second trait, say butterfat concentration, and so on. The snag here is that traits such as these two may be negatively correlated, i.e. by switching to selection for butterfat one may lose some of the accumulated improvement in milk yield. This form of selection applies to both bulls and cows.

(2) *Independent culling levels.* In herds where it is possible to cull cows for production weaknesses rather than diseases such as infertility, mastitis and lameness, acceptable standards can be set for a number of traits such as (a) yield of fat plus protein, (b) conformation of legs, (c) conformation of udder. Any cow failing to meet minimum standards for any of these traits should be culled.

(3) *Index selection.* Animal breeders like to combine different traits into a single genetic index which weights the individual components of the index according to their economic importance. The Milk Marketing Board currently proposes a *Profit Index* which weights the *estimated transmitting abilities* (ETA) of a progeny tested bull as follows:

Profit index = + 0.94 (fat ETA kg) + 2.75 (Protein ETA kg) – 0.039 (Milk ETA kg)

This index recognises that increased profit is best achieved by an absolute increase in yield of especially protein and, to a lesser extent, fat achieved by increasing the concentration of these constituents in milk with a marginal reduction in total milk production.

Selection in practice

Dairy bulls

The widespread use of artificial insemination makes it possible for a dairy farmer to select dairy bulls for breeding whose progeny have been tested and shown to be superior to their contemporaries. Suppose a farmer with 100 milch cows wishes to replace 25% of his herd per annum. He mates the better milking half of his herd to one or more superior dairy bulls and gets, on average, 25 potentially superior heifers and 25 rather inferior dairy 'type' bull calves. The other half of the herd, and all the heifers, will probably be mated to beef bulls.

Details of bulls in breeders' catalogues include a picture of the bull, usually standing uphill or with the camera tilted to one side to give an impressive-looking topline, pictures of some of his daughters and (far more importantly) detailed information as to:

(1) Yields of milk solids by his daughters relative to their contemporaries (the *Improved Contemporary Comparison*, ICC).
(2) Daughter conformation.
(3) Daughter sizes, temperament, ease of milking and calving difficulties.
(4) Pedigree of bull:
  (a) performance records of dam and granddam;
  (b) performance of female progeny of sire and grandsire.

The records of two imaginary stud bulls are given in Table 11.5. Although the descriptions follow the conventions of MMB, any resemblance to a real bull, dead or alive, is coincidental.

The lactation performance of each bull's daughters is currently assessed by the Improved Contemporary Comparison (ICC) procedure from a baseline of 1983. Full details of the ICC are given in ADAS Booklet 2405 (1982). In essence, thousands of Friesian/Holstein bulls are mated with heifers right across the UK dairy herd so as to rank them according to the performance of their heifer progeny against a

Table 11.5 Performance, conformation and other characteristics of two (imaginary) premium Friesian/Holstein dairy bulls

	Gwernyfed Charger	Canterbury Magnum
ICC(s) 83 January 1991		
Milk	+570	+69
Fat	+22.9 +0.03%	+11.9 +0.23%
Protein	+15.2 –0.02%	+10.6 +0.10%
Daughter distribution,		
herds	53	63
daughters	166 (wt.127.2)	85 (wt.75.2)
Profit index	£41	£38
Daughter conformation (difference from breed average)		
Stature, small-tall	+3.2	+0.4
Chest, narrow-wide	0.0	+0.7
Body depth, shallow-deep	+3.6	+1.6
Angularity, thick-sharp	+3.2	+0.1
Rump, highpins-lowpins	–0.8	+0.3
Rump width, narrow-wide	+0.4	+2.4
Rear legs side, straight-sickled	+0.2	+0.9
Rear legs rear, close hocks-straight	–1.2	+0.1
Foot angle, low-steep	–0.4	–1.4
Fore udder angle, steep-level	+0.2	0.0
Fore udder attachment, weak-strong	+0.4	–1.2
Rear udder width, narrow-wide	+2.3	–0.2
Udder support, broken-strong	+1.2	–0.1
Udder depth, deep-shallow	0.0	–1.6
Teat placement, rear; wide-close	–0.5	–0.2
Teat placement side; close-wide	+1.4	–0.5
Teat length, short-long	+0.2	+0.8
Beef shape, poor-good	–3.2	+0.4
Temperament, flighty-quiet	+1.3	–0.8
Ease of milking	+1.6	+0.2
General characteristics		
Calving difficulties when used on Friesian/Holstein cows (%)	4.3	1.7
Average gestation length when used on Friesian/Holstein cows (days)	285.3	281.2

fixed genetic base. First lactation records of daughters of bulls on test and those of their contemporaries are standardised by correcting for age at calving and season of calving. Other adjustments are made to eliminate bias attributable to the small amount of annual progress in the progeny of groups of bulls tested in successive years and to minimise bias due to non-random matings, i.e. the possibility that a particular bull may appear superior through having been mated to a superior group of cows.

The ICC figures in Table 11.5 predict the average merit of all future daughters of each imaginary bull for milk yield and composition relative to a fixed baseline which is currently the average performance of Friesian-Holstein heifers tested in 1982/83. Table 11.6 compares these values with those for actual bulls within the Genus catalogue of proven sires (1991/92) and average ICC 83 values for all Friesian/ Holstein bulls tested by MMB to 1990. The average of all bulls tested tends to give negative values for ICC 83. The best bulls, however, which are selected as proven sires have strong to very strong positive values for total milk yield and fat yield (or concentration). Protein yields tend to follow milk yield. The scope for variation in protein concentration by conventional breeding is very small. Relative to other proven sires, Gwernyfed Charger is an outstanding sire with respect to his capacity to increase total yield of milk solids in his daughter. Canterbury Magnum is a very different sort of sire. His capacity to increase total milk yield is relatively modest but he can improve both fat and protein concentration, the latter perhaps to a rather fanciful degree. His Profit Index is therefore almost as great as that of Gwernyfed Charger. Which bull would be superior for a particular

Table 11.6 Genetic merit of Genus Proven Sires (1991/92) compared with average values for all tested MMB bulls born in 1971 and 1981

| ICC 83 values | Proven sires | | Average values for bulls born in | |
	Mean	Range	1971	1981
Milk (kg)	+269	+15 to +616	−116	−64
Fat (kg)	+12.7	+4.4 to 24.4	−5.0	+0.9
Protein (kg)	+9.2	+0.9 to +15.0	−3.2	−1.2
Fat (%)	+0.07	−0.13 to +0.30	−0.01	+0.10
Protein (%)	0.0	−0.08 to +0.07	−0.01	0.0

farm would depend on the economic consequences of improving milk yield or milk concentration in their particular circumstances.

The reliability of these predictors of the merit of each bull is obviously governed by the size and distribution of the sample of daughters involved in his progeny test. The 'weighting' value given with each ICC is obtained by a simple arithmetic gambit. For each herd in which the bull is tested a weighting value is obtained by comparing the product and the sum of the number of daughters (*nd*) and contemporaries (*nc*). Thus if in a particular herd a bull is tested on 4 daughters and there are 12 contemporaries, the bull's weighting for that herd is:

$$\frac{nd \times nc}{nd + nc} = \frac{4 \times 12}{4 + 12} = 3.0$$

The overall weighting for each bull is the sum of its individual weightings for each herd in which it is tested. The reliability of the ICC increases sharply at first but progressively levels off to reach a plateau. For a bull with an ICC fat yield of +15 kg and a weighting of 10, 20 or 200, one would predict that the improvement achieved by 95% of his daughters would be within the following ranges:

Weighting	95% range (kg)
10	+3.4 to +26.6
20	+8.1 to +21.9
200	+13.0 to +17.0

In practice, any ICC figures with a weighting over 60 may be considered reliable enough.

The use of a fixed baseline (currently 1983) for the ICC has two practical advantages:

(1) Farmers can directly compare the relative merit of old and new bulls.
(2) A bull's ICC will tend to remain reasonably constant and thus provides a true picture of his genetic merit even though tested over several years.

However, the actual improvement achieved by use of a nominated bull in a particular herd may be less than that predicted for his ICC if the average genetic merit of the cows to which he is mated (a) was significantly above the Friesian/Holstein mean at the time when the 1983 baseline was established, or (b) if significant genetic progress has been achieved since 1983. Thus the greater the genetic merit of the cows relative to the 1983 baseline, the further the actual gains achieved by the bull will fall below those predicted from his ICC (83). The previous baseline for the ICC had been set in 1974. By 1983 the *average* ICCs of the sires of Friesian/Holstein heifers was +220 kg milk, +10.2 kg fat and +6.4 kg protein, against the 1974 baseline. Any bull with ICC (74) figures below this was below the average for the breed.

The MMB Panel and Livestock Officers also assess the conformation of daughters of their nominated bulls and bulls on test. Each heifer is given a score from 1 to 9 for 17 points of conformation. Table 11.5 lists these points and indicates the extremes described by 1 and 9 respectively, with average values for Friesian/Holsteins being between 4.2 to 5.2.

For example:

Stature, or height at withers:
 1 = very small (<125 cm); 9 = very tall (>149 cm).
Rear legs, rear view:
 1 = very close hocks with severe toe out; 9 = straight.
Rear udder width:
 1 = very narrow; 9 = very wide.
Udder support (suspensory ligament):
 1 = no cleft, broken support; 9 = extreme cleft, strong support.

For characters such as stature, chest width, angularity, etc. the ranking 1–9 is merely descriptive and carries no clear message as to what is good or bad. It is of little consequence if cows that produce more milk have longer legs unless, for example, they become too long for the cubicles. However, for characters such as beef shape, udder width, depth and support and rear legs (rear view), the higher the score the better.

Gwernyfed Charger (Table 11.5) now stands revealed as the dairy farmer's image of the typical Holstein sire. His daughters are milky, tall and angular, with better than average udders but very poor beef shape. The daughters of Canterbury Magnum are close to the Friesian–Holstein average on most points of conformation but with wide rumps and slightly above average beef shape. When used on Friesian/Holstein cows this bull has contributed to 1.7% serious calving difficulties. Gwernyfed Charger with 4.3% should be restricted to cows with proven records of uncomplicated calvings.

The relative merits of Friesian and Holstein cattle (or at least our conventional view of what constitutes a Friesian or a Holstein) is a subject that arouses passionate debate and no little acrimony. The argument really arises from the nature of correlated responses to selection, i.e. the impossibility of selecting for improvements in all desirable traits simultaneously. Pure Holstein sires have undoubtedly contributed greatly to the genetic improvement of the British black and white dairy herd both in terms of milk yield and composition. They have also tended to produce heifers with 'better' udders, shallower, wider and more strongly supported. However, presumably in conse- quences, this has tended to make them more 'cow hocked' – i.e. closer hocks with toes turned further out. The implication of this is that selection for good udders may be incompatible with selection for good legs, or, to be more commercial, selecting with the intention of reducing environmental mastitis might increase the incidence of lameness or *vice versa*. I notice that in about half of the bulls in the 1991/92 Genus catalogue, the Linear Assessment does not include a value for rear legs, rear view to indicate conformation within the range close hocks to straight legs. I wonder why?

Cow selection

Unless a dairy farmer is in the business of breeding cows as potential mothers of premium bulls, the selection decisions that he can profitably make with regard to the cows in his own herd are rather limited. He can:

(1) choose to cull individual cows on grounds of inadequate produc- tion or poor conformation;
(2) select the bull most appropriate to each dairy cow on the basis of her performance, conformation and ease of calving.

If he is in the happy position of having so few casualties due to infertility, mastitis or lameness that he can cull on grounds of poor performance, the best approach is probably that of setting an independent culling level for fat plus protein yield over 2–3 lactations. Any cow that is in, say, the bottom 20% for the herd is culled.

It would also be nice to cull on poor conformation, especially with regard to legs and udder. However, this is, in practice, likely to be an unrealistic luxury. Moreover, there is nothing inherently wrong with poor shape *per se* unless it predisposes to problems such as mastitis and lameness, which are likely to lead to early culling anyway.

The Milk Marketing Board provide two useful indices of cow

performance based on National Milk Records. The *Cow Production Index* marks individual cows within a herd on the basis of their yield of milk solids in the most recent lactation, reduced to a common first lactation equivalent after adjustment for lactation numbers, age and season of calving. Each cow is compared against a base of 100. This gives the farmer a little more information than that which he can get from simple records of total lactation yield, but not much. The *Cow Genetic Index* (CGI) marks each cow according to her genetic potential for fat plus protein production and is based on her production in lactations 1–5, the ICC of her sire, the CGI of her dam and the average genetic merit of the herd.

The CGI is calculated from the individual fat and protein yield indices, thus

$$CGI = 10 \times (\text{Fat index} + \text{protein index}) + 500$$

The base value of CGI = 500 for a cow is equivalent to a bull with an ICC for fat plus protein of zero. The CGI figure is a prediction of the weight of fat plus protein yield that a cow will transmit to her progeny, relative to one with a CGI of 500. Thus a cow with a CGI of 700 will transmit 20 kg of fat plus protein ($10 \times 20 + 500$). The CGI ranks individual cows irrespective of the herd in which they are milked and the feeding programme for that herd. It is therefore a particularly useful indicator of the genetic value of a dairy cow as a mother of beef bulls or as a donor in a MOET scheme.

Nutrition-genotype interactions

The genetic improvement schemes described above are designed to be as independent as possible of environmental effects on performance, principally those attributable to differences in nutrition. Moreover, differences in the nutrition of the UK dairy herd are not perhaps so great as to distort the ranking of sires and dams in terms of ICC and CGI even if actual performance of their offspring in a particular herd does not match expectation. Nevertheless it is useful to examine what are the effects of selection for performance on high energy diets when transposed to a feeding system far more dependent on forage. Table 11.7 presents very recent results from the Langhill Herd of the Scottish Agriculture College (Oldham, Simms and Marsden, 1992). Cows with CGI values of 700 and 500 were fed high concentrate (0.45 total DM) and low concentrate (0.20 total DM) rations, the remainder being

Table 11.7 Performance of Friesian/Holstein cows with high and average genetic indices on high (0.45) and low (0.20) concentrate rations (Oldham, Simms and Marsden, 1992)

	High concentrate			Low concentrate		
Predicted CGI	700	500	700:500	700	500	700:500
Milk yield	7032	6164	1.14	6030	5309	1.13
Fat (g/kg)	42.5	44.7	0.94	46.2	45.1	1.02
Protein (g/kg)	30.4	31.5	0.96	30.4	31.3	0.97
Fat plus protein (kg)	512	466	1.10	463	406	1.14
Dry matter intake (kg)	4645	4408	1.05	4065	4010	1.01
Efficiency (Milk MJ/MJ ME)	0.388	0.366	1.06	0.414	0.367	1.12
Margin overall food costs (£)	759	643	1.18	732	599	1.22

provided by forage. The relative increases in total milk yield and in yield of fat plus protein were very similar on both rations. Milk fat concentration tended to be greater on the low concentrate ration for both groups of cows. The high CGI cows tended to have a greater dry matter intake on the high concentrates ration only. Efficiency of ME conversion to milk was, in fact, highest for the high CGI cows on the high forage ration although (inevitably) the high CGI, high concentrate cows made most profit, not only in relation to feed costs but even more so if fixed costs are taken into account.

This is only one set of results (although a powerful set) and it would be interesting to examine greater variations in nutrient supply. Nevertheless it does suggest that genetically superior cattle as defined by ICC and CGI will express their superiority throughout the realistic range of diets fed to UK cattle. Moreover, this response to a proper programme of genetic selection is likely to persist at a commercially realistic rate of 1–2% per annum for the foreseeable future. This annual rate does not sound too impressive but after 20 years the accumulative effect would look very good.

I cannot leave this chapter on improvement of dairy cow performance by breeding without reference to figures I obtained of individual cow performance in one of the top MMB recorded herds owned by Mr G.H. Coaker at Kegworth. When expanding his herd he selected cows and heifers on their looks and their performance to date – nothing more subtle than that. After arrival they were milked three times daily and fed a very high energy ration. Average yield for the herd in 1987 was 9919 kg at 4.05% fat and 3.18% protein. The performance of cows in second and third lactation was as follows:

	Lactation		
	2nd	3rd	3:2(%)
Home-bred cows	8501	9152	+7.6
Cows bought in after second lactation	7390	9566	+29.4

The increase of 7.6% in home-bred cows can be attributed to increasing maturity. The increase in cows purchased after their second lactation was 29.4%, implying that 21.8% (29.4 - 7.6) could be attributed to changes in feeding and management.

Genetic improvement is real but slow and must be put into true perspective alongside the potential for change through nutrition and management. This is particularly important since most of the 'literature' related to animal breeding that lands on a farmer's desk is produced by companies who have something to sell – breeding stock, semen or ova – and so have a vested interest in persuading their customers that their products are a sure-fire way of making more money.

Beef bulls

The choice of a beef bull for dairy cows is an altogether simpler matter. The specialist dairy farmer wants a live calf that can be sold into a beef unit at a good price and an uncomplicated calving which does not compromise the cow's lactation or her ability to return to service.

The first criterion of a good beef bull for dairy cows is that he does not carry a high risk of calving difficulties. Table 3.1 showed that Hereford and Aberdeen Angus bulls cause few calving problems. Charolais bulls, on average, are more likely to cause problems by virtue of their size, shape and the fact that they prolong pregnancy (284 days for a Charolais/Friesian compared with 281 days for a pure Friesian). The Limousin bull extends pregnancy even longer (287 days) but the Limousin/Friesian calf is smaller, on average, than the Charolais/Friesian calf, so is less likely to cause problems at birth. There are, however, wide differences between bulls within breeds in 'ease of calving'. A dairy farmer requesting semen from a Limousin or

Charolais bull would normally select one with a good record in this respect.

In recent years there has been a great increase in the use of 'double-muscled' Belgian Blue bulls to inseminate dairy cows. The gene which is responsible for this condition, which has appeared in most beef herds, does not, in fact, alter the number of muscles but predisposes to abnormally overdeveloped muscles, especially in the expensive cuts around the hind quarters. The attraction of such animals to the butcher is obvious. However, this excessive development of the hindquarters is associated with a relative reduction in the dimensions of the pelvic canal to the extent where normal calving becomes extremely difficult or impossible. In many purebred herds of Belgian Blue cattle it is normal to deliver all calves by premeditated Caesarian section performed under local anaesthesia. It is the view of the UK Farm Animal Welfare Council that this premeditated, repeated imposition of major surgery to overcome a deliberate and potentially painful distortion of normal shape constitutes an unacceptable insult to welfare. I must stress however that the obstetric problem is predominantly a consequence of the distorted shape of the double-muscled cow rather than the size of her calf. The incidence of obstetric problems in Friesian cows inseminated by Belgian Blue bulls appears on average to be no worse than for other large breeds such as the Charolais.

Many dairy farmers sell their bull calves within the first two weeks of life, in which case their income is determined overwhelmingly by the breed and type of the bull. Current calf prices (1992) in the South of England are:

	Bull (£s)		Heifer (£s)	
	Average	Range	Average	Range
Hereford × Friesian	135	70–180	90	40–120
Friesian/Holstein	130	30–180	–	
Charolais	205	140–260	130	70–170
Limousin	185	70–230	110	70–140
Simmenthal	190	100–240	135	60–190
Belgian Blue	215	140–280	185	120–240

Top prices are being paid for Charolais × and Belgian Blue × calves. At the height of its fashion many Belgian Blue calves were fetching

£300, a price which grossly inflated their genetic merit. When Limousins were in the height of fashion their price too became inflated. Current prices reflect, I believe, a better-informed market. The extraordinary range of £30–180 in prices for black and white calves reflects the real difference in size and confirmation between a strong calf of Friesian type and the worst sort of Holstein.

If a dairy farmer can make over £200 for a Charolais or Belgian Blue × calf at two weeks but only £135 for a Hereford × Friesian it is difficult to persuade him that the Hereford × might indeed be as efficient a producer of beef as the Charolais. His economic objectives are simple: to get the best possible calf price without significantly increasing the risk of obstetric problems that might (at worst) kill cow or calf or (more likely) impair her subsequent lactation and fertility. He is therefore likely to select beef bulls with a low incidence of obstetric problems (perhaps Herefords, Aberdeen Angus or Limousins) for his heifers and for any cow with a history of calving difficulties. Big, broad cows with a record of easy calvings may as well be sired by a Charolais. Use of a double-muscled Belgian Blue implies acceptance of the way in which that purebred bull was produced.

If the dairy farmer intends to rear his own beef calves then his choice of sire needs to be rather more educated. Effects of sire breed on the efficiency of growth in beef cattle are complex and outside the scope of this book (see Allen, 1990; Webster, 1989). There are however two crucial points:

(1) *Breed differences in the biological efficiency of growth are very small.* In Chapter 1 I illustrated the close similarity between Friesian and Jersey cows, goats and bitches in the biological efficiency of lactation (Table 1.6) and, in this chapter, have already made the point that genetic variation within species in matters of vital importance for survival, like the efficiency of utilisation of nutrients for growth and lactation tends to be quite small. Table 11.8 compares growth rates, carcass yield and the efficiency of utilisation of food for gain by beef cows from Friesian cows by four sire breeds. As expected, Charolais × calves were the largest with Aberdeen Angus and Hereford × calves smallest and Limousin × calves intermediate. The offspring of the British breeds had a little more fat trim but not much. The continentals had better killing-out percentages and slightly more saleable meat in the carcass. However, on this semi-intensive system the British breeds tended to have slightly better food conversion efficiencies. As a general

Table 11.8   Performance of calves from Friesian dams (from Southgate, 1982)

| | Sire breed | | | |
	Angus	Hereford	Charolais	Limousin
Weight at slaughter (kg)	363	410	494	454
Food conversion ratio				
(g gain/kg feed)	86	88	82	85
Killing-out (%)	52.5	52.3	54.8	54.7
Saleable meat in carcass (%)	72.5	71.9	72.7	73.3
Saleable meat in expensive				
cuts (%)	44.1	44.1	44.8	45.4
Fat trim in carcass (%)	9.6	9.7	9.0	9.2

and oversimplistic rule, calves from Continental bulls tend to be superior when fed a high nutrient density diet that allows them to express their full potential for lean tissue growth within the constraints of their capacity to consume dry matter. As the diet becomes progressively more fibrous (lower M/D) and the system more extensive the traits conferred by Aberdeen Angus and Hereford bulls become progressively more attractive. The pure 'double-muscled' animal, whether Belgian Blue or any other beef breed is not, in fact, particularly efficient at growth. Even if they stay healthy (and they are particularly prone to diseases like pneumonia) they tend to have poor appetites and are difficult to finish. This is an extreme example of the fact that correlated responses to selection are not always favourable. Selection of an animal for extreme leanness, high muscularity and high killing-out percentage tends to select for a relatively small gut and thus a low appetite.

(2) *Effects of sire breed on gross profit margin are almost entirely the same as differences in price at birth.* Table 11.9, taken from the *Meat and Livestock Commission Beef Yearbook* (1990), presents figures for physical and financial performance for producers rearing calves from the dairy herd for beef on two intensive systems, silage + concentrates and cereal or 'barley beef'. I think we may assume that the calves in this trial were more typically Friesian than Holstein. Differences in growth rate between the pure Friesians

Table 11.9   Physical and financial performance of entire male calves from Friesian cows in two intensive beef systems (MLC, 1990)

	Intensive silage/concentrate			Cereal beef	
Sire breed	Friesian/ Holstein	Charolais	Limousin	Friesian/ Holstein	Continental cross
Physical performance					
feeding period (days)	456	412	444	305	308
slaughter weight (kg)	531	560	517	483	494
LW gain (kg/day)	1.01	1.06	0.98	1.31	1.34
concentrate feed (tonnes)	1.35	1.26	1.20	1.99	1.83
silage (tonnes)	6.1	7.1	6.3	–	–
Financial (£/head)					
sales	603	698	626	564	610
less calf & mortality	168	283	202	193	253
variable costs, feed	169	161	156	260	247
other	66	64	50	29	24
Gross margin per head	200	190	218	82	86

and the continental crosses were quite small. The Charolais ✕ calves finished most rapidly and at the highest weights. The Limousin ✕ were the cheapest to feed. However, the gross profit margins per head within each system were very similar because the financial advantage conferred by the beef sire through improved growth rate and food conversion efficiency were all taken up by the differences in the price of calf at purchase. Once again this makes the point that the practical advantages to the commercial farmer of buying genetically superior stock are often oversold. When Belgian Blue ✕ Friesian calves were fetching £300 per head I could not blame dairy farmers for breeding them and selling them at birth. I would not, however, have been so daft as to buy one.

12 Fertility

The dairy cow is expected to deliver one calf per year without problems at the time of year deemed most appropriate by the farmer, while sustaining a lactation of over 300 days. For a number of reasons which will be discussed in turn, it is generally agreed that the optimal *average* interval between calvings for the herd should be 365 days. To achieve this, all aspects of the reproductive cycle have to proceed normally and without interruption.

(1) After calving, the uterus must be free from infection and regress rapidly to the non-pregnant state.
(2) The ovaries must recommence the oestrous cycle of follicular development, ovulation and formation of a corpus luteum.
(3) The cow must display sexual behaviour which is recognised and acted upon (rarely) by a fertile bull, or (commonly) by the stockman who then arranges for the cow to be artificially inseminated.
(4) The single artificial insemination must achieve a successful fertilisation.
(5) The fertilised ovum must implant in the endometrium of the uterus and avoid the range of hazards that can produce early foetal death or abortion in late pregnancy.
(6) Mother and calf must emerge unharmed from the ordeal of parturition.

All this is a lot to ask.

Any extension of the average calving interval costs the farmer money. In 1985, Esslemont, Bailie and Cooper calculated that every day's increase in calving interval beyond 365 days costs £1.36 per cow, this being the difference between the decrease in milk sales and food costs. They added a further loss of 45p/day for autumn-calving herds due to cows calving later in the winter and so producing more milk when the price per litre is lower, giving a total loss of £1.81 per cow-day.

Any cow that had to be culled for infertility was calculated to cost a further £415, this being the difference between her cull value as cow beef and the cost of bringing a new heifer into the herd. Table 12.1 uses these figures to derive an estimate of the economic importance of failure to meet the fertility target of 365 days between successive calvings.

Table 12.1 Effect of fertility management on dairy herd profitability (from Esslemont, Bailie and Cooper, 1985)

	Good herd	Poor herd	Difference
Heat detection (%)	80	50	30
Conception rate (%)	60	50	10
Mean, calving to first service (days)	65	85	20
Calving to conception (days)	81	107	26
Cows culled for infertility (%)	1	12	11
Cost difference per cow			
delayed conception	(26 × £1.81)	45.06	
culled for infertility	(0.11 × £415)	45.65	
semen costs		3.85	
extra veterinary costs		6.00	
	Total	102.56	

The 'good herd' is achieving an average calving-to-conception interval of 81 days (given a calving interval of 362–365 days) by virtue of a heat detection rate of 80% and a conception rate (to AI) of 60%. The 'poor herd' has an average calving-to-conception interval of 107 days (calving interval of 389–392 days) and 12% of the cows are culled for infertility. This level of performance, which is poor but by no means unusual, represents an annual loss of about £100 per cow relative to the herd with a calving interval of 365 days and minimal culling for infertility.

Esslemont's calculations assume that the costs of infertility increase linearly with increasing average calving interval for the herd. It would be a mistake, however, to interpret this as meaning that every cow whose calving interval slips beyond 365 days is losing the farmer money. A calving-to-conception interval of 70–100 days giving a calving interval of 350–380 d for individual cows (with a mean of 365 d) may be considered optimal. Any cow that fails to be served within

120 days of calving or, worse, is served but discovered to be non-pregnant, say, 150 days after calving, is not earning her keep – though it may not be her fault.

Fertility management

The management of dairy cows and heifers to ensure optimum fertility depends, at the outset, on the proper recognition of oestrus. This is one of the simplest aspects of fertility control in theory yet the one that tends to be the least successful in practice.

Oestrous behaviour

The oestrous behaviour of cows is bisexual and indiscriminate. When a small group of feral cows runs with one or more bulls at a ratio of about 10 cows per bull, the bull identifies cows about 2 days before the onset of oestrus or standing heat and tends to isolate them from other females. Bull and cow may favour each other's company for 2–3 days. However, the period of oestrus when the cow will stand to be mated is only about 16 hours (range 8–24 hr). The number of copulations, as distinct from unsuccessful mounts, may be 2–4.

When cows are at pasture in the absence of a bull or when the ratio of cows to bulls is large (more than 40:1), the cow that proceeds through pro-oestrus into oestrus displays a number of typical signs of increasing sexual arousal and furthermore becomes sexually attractive to other cows in the herd. She is said to be on heat or 'bulling'. In pro-oestrus the cow becomes progressively restless. She may approach other cows in oestrus, sniff and lick around the area of the vulva and attempt to mount them. More significantly, she becomes increasingly attractive to other cows and increasingly tolerant of their attentions. In pro-oestrus she will tolerate licking and sniffing but move away if another cow attempts to mount. In oestrus she will stand to be mounted from the rear (Fig. 12.1), hence 'standing heat'. This is the most reliable single behavioural indication of a cow in oestrus.

An individual cow is more likely to make active sexual approaches – sniffing, licking and mounting – when in oestrus or pro-oestrus but may perform these actions at any stage of the oestrous cycle or when pregnant. It follows that the cow making the approach is probably *not* in oestrus. There is one exception to this. Occasionally one cow will

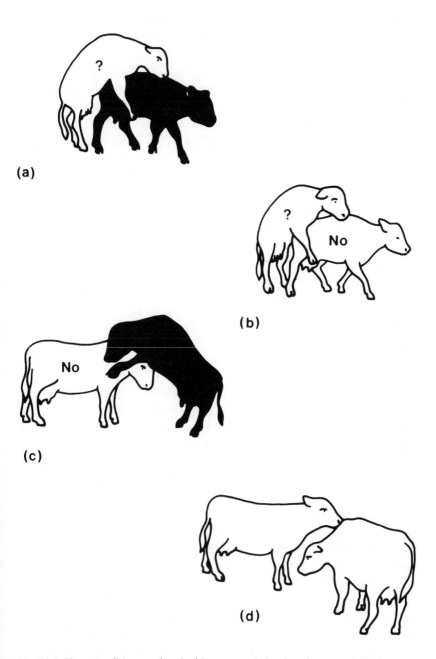

Fig. 12.1 Signs possibly associated with oestrous behaviour in cows. Only the cows marked in black are in oestrus. (a) Cow in oestrus stands to be mounted. (b) Cow not in oestrus rejects attempt at mounting. (c) Cow in oestrus mounts head to head. (d) Mutual grooming, one or other cow *may* be in pro-oestrus.

mount another head to head (Fig. 12.1). In this case the cow doing the mounting is probably, but not certainly, in oestrus. The one responding to this advance head first is definitely not.

Other indicators of oestrus include signs of rubbing around the tailhead, a temporary drop in milk yield and a disturbance to the cow's normally placid temperament. She may lose interest in food and, if at pasture, stand at a hedge all day bawling for a bull. If in winter housing she may fail to take her customary place in the queue for the milking parlour.

Table 12.2   Oestrous behaviour of cows in groups, housed over winter and at pasture over summer

	Winter housed	Summer at pasture
Duration of pro-oestrus and oestrus (hours)	12–24	24–36
Duration of oestrus (standing heat, hours)	8–16	16–28
Number of mounts	40–80	80–120

Table 12.2 presents some very approximate figures relating to oestrous behaviour in cows. There is no doubt that the intensity and the duration of oestrous behaviour in cows at summer pasture are both considerably greater than in cows in a cubicle house during the winter. This is, in part, due to decreased secretion of oestrogen during the dark winter months but can equally be attributed to the separation imposed by cubicles and the cows' reluctance to mount one another on wet, slippery floors. It is generally recognised that detection of oestrus is easier when cows are group-housed in strawed cattle courts.

The number of times a cow allows herself to be mounted or mounts another while on heat is really very large: 40–120 episodes. In summer, it is most unlikely that a herdsman would fail to notice a cow 'bulling' while bringing her in from pasture to the milking parlour. However, in winter, when the number of episodes is less and may tend to occur during the night when the cows are otherwise undisturbed, the chances of a cowman making an accurate diagnosis of oestrus during a period of observation lasting 15 min. is probably less than 50%. Esslemont has investigated the relative effectiveness of various strategies for detecting oestrus from cow behaviour. Two 15-min. periods of observation associated with morning and evening milking achieve a detection rate of only 55–65%. The most effective strategy appears to be three periods of observation of at least 20 minutes spaced as evenly as possible throughout the 24 hours, say at 0600, 1400 and 2200 hrs. This achieved a detection rate of 84%. The final, night time round at 2200 (10pm) appears to be particularly effective for housed cows.

Aids to oestrous detection

The two first essentials for accurate heat detection are to know which cows to look out for and to recognise these when you see them. Each day the herdsman should be able to check easily from wall calendar or computer print-out:

(1) Cows due to come into oestrus for the first time (i.e. those which calved more than 15 days ago but have not previously shown oestrus).
(2) Cows due for service (i.e. calved more than 45 days but as yet unserved).
(3) Cows which last showed oestrus 18–23 days ago and were or were not served.
(4) Cows which have already been served more than once, or served but diagnosed as non-pregnant.

The farmer or herdsman with a small dairy herd can undoubtedly recognise each of his cows from their natural physical characteristics. However, to ensure that cows in a large herd at pasture or in the cow house can be reliably identified by farmer, herdsman or relief milker, it is necessary to mark them clearly, preferably at both ends, by (for example) a freeze-brand and a large ear tag.

Other aids to heat detection include:

(1) *Colour marking* the area of the tailhead with a dyed paste to indicate that the cow has been mounted. The Kamar Heat Mount Detector works on this principle. The dye is contained within a plastic phial which is stuck to the tailhead. Mounting causes the dye to be squeezed out and change colour when it contacts the skin of the cow. (Occasionally, mounting merely causes the heat detector to fall off.)
(2) A coloured *chin ball* worn by a bull running with cows or heifers, or by a vasectomised 'teaser' bull, or possibly a cow treated with

testosterone to exaggerate her natural behaviour to cows in oestrus, can be extremely effective and is certainly a practical consideration for herds with a poor calving interval due to poor heat detection by the herdsman or team of contract milkers.

(3) *Swollen, reddened appearance of the vulva accompanied by mucus and perhaps streaks of blood.* This sign tends to be most prominent in late oestrus. While it can be helpful, herdsmen who tend to rely on it will probably inseminate their cows too late.

(4) *Progesterone concentration in milk.* The secretion of progesterone from the corpus luteum is high during dioestrus (see Fig. 3.3) and increases further during pregnancy. Minute quantities of this hormone enter the milk which can be analysed as an aid to the detection of oestrus or pregnancy. The milk progesterone assay which can be done by the farmer acting on veterinary advice is probably most useful as an early test for pregnancy (q.v.). However, milk progesterone levels in a non-pregnant cow fall sharply about 19 days after previous oestrus and remain low for about 4 days through pro-oestrus. Analysis of milk samples for progesterone on Days 7, 19 and 21 can be a good predictor of the time of oestrus. It does, however, presume that the operator correctly identified the time of previous oestrus. Even then, it is not foolproof and not really a practical substitute for diligent observation of oestrous behaviour. Its best application is probably for cows that have already been inseminated at least once. Milk progesterone assays at 2-day intervals from Day 18 to Day 24 after previous service should identify any cow that has failed to conceive in time for her to be served at her next oestrus and will furthermore give a reasonably reliable early diagnosis of pregnancy by 24 days (Ball, 1983).

Synchronisation of oestrus

One way to get round the problem of failure to recognise oestrus is to control the time at which it occurs. There are two ways of doing this, both based on controlling the time of regression of the corpus luteum and so the end of dioestrus and the beginning of pro-oestrus.

PROSTAGLANDIN

A single injection of prostaglandin (PG) will destroy the corpus luteum and so bring cows that were previously in dioestrus into standing heat in 3–4 days. PG is only effective over the last 10–12 days of dioestrus when the corpus luteum is actively secreting progesterone. Injection of a group of heifers on a single day would therefore only bring half of them into heat together. PG will not affect a cow that is in anoestrus (i.e. one whose ovaries are inactive and is not 'cycling'). If injected into a pregnant cow it is likely to induce abortion. It therefore needs to be used with care under strict veterinary supervision. The options are:

(1) Two injections of PG at an 11-day interval. It is assumed that whatever the stage of the oestrous cycle at the time of the first injection, all animals should be in dioestrus at the time of the second. Animals may then be inseminated, probably twice, 3 and 4 days after the second injection. The technique is appropriate for the farmer who wishes to use artificial insemination (AI) on maiden heifers running at pasture.

(2) Identification of a corpus luteum in a non-pregnant cow by rectal examination, followed by a single PG injection, heat detection and one AI (or two AIs on successive days if heat detection is very poor). This approach is obviously more suited to adult cows already in the dairy herd.

PROGESTERONE

The alternative to synchronising the time for active destruction of the corpus luteum by prostaglandin is to inhibit the development of the next follicle by the use of progesterone. The most effective way to achieve this is to insert into the vagina a PRID (progesterone-releasing intravaginal device). This is withdrawn after 12 days, the inhibitory effect of progesterone on the release of follicle-stimulating hormone is removed, and the cow comes into oestrus within 2–3 days. In addition to progesterone, PRIDs contain some oestrogen to aid in the destruction of the original corpus luteum. The mix of oestrogen and progesterone in PRIDs is also usually effective in stimulating follicle development and ovulation in cows previously in anoestrus with inactive ovaries.

The probability of fertilisation from a single AI following synchronisation of oestrus by prostaglandin or progesterone is slightly less than that which occurs following a normal oestrous cycle. Table 12.3 (from Blowey, 1985) compares the effectiveness of observation and synchronisation followed by AI in achieving fertilisation. The percentage number of cows pregnant after three weeks is 52 for synchronisation compared with 48 for herds with good heat detection; in effect, no real difference.

Table 12.3   The relative effectiveness of observation and synchronisation in achieving fertilisation following a single insemination by AI (from Blowey, 1985)

	Heat detection rate (%)	Conception rate (%)	Cows pregnant after 3 weeks (%)
Observation and AI:			
good heat detection	80	60	48
poor heat detection	50	60	30
Synchronisation	95	55	52

Synchronisation of oestrus is only likely to be worth the expense of time and money if:

(1) current rates of heat detection are very poor.
(2) there are sound management reasons for batch-calving, e.g. to ensure that heifers enter the dairy herd at the most cost-effective time of the year.

The use of drugs as a remedy for poor heat detection smacks of bad husbandry – an expensive substitute for the gentle business of watching cows. Synchronisation of oestrus to control the time of calving for heifers can be useful provided it is borne in mind that only half the heifers will conceive first time round. Those that conceive at the second and third cycle after synchronisation will calve progressively out of synchrony due to natural variations in the length of the oestrous cycle. Furthermore, no farmer should embark upon a programme for synchronisation of oestrus without first ensuring that he has adequate facilities for batch calving and can maintain adequate hygiene in calving boxes that are in constant use (Chapter 8).

Time of insemination

Cows normally ovulate about 12 hours after the end of standing heat. The optimal time for insemination is 12–18 hours before ovulation, i.e. during the last six hours of oestrus. Thereafter the probability of

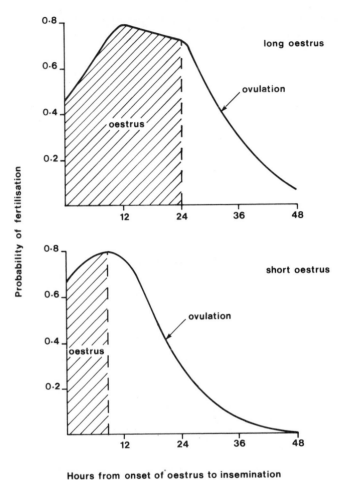

Fig. 12.2 Effect of time of insemination on fertility of cows with a long and short period of standing heat.

fertilisation declines sharply (Fig. 12.2). Assume a cow is seen bulling for the first time prior to morning milking at 0600 and that the herdsman has observed the very first time that she stands to be mounted right at the beginning of oestrus. He then rings the AI service who inseminate the cow at 1000. Assume further that she has a long oestrous period of 24 hours. In this extreme situation she is inseminated four hours after the onset of standing heat and about 32 hours before ovulation. Fertility may be sub-optimal, about 60%, but no worse than if insemination were delayed to the following day (Fig. 12.2). If she had a short heat of 8 hours, insemination on the following day would carry a low probability of fertility (less than 40%). It follows that cows seen bulling in the morning should be inseminated the same day, especially during the winter months when the duration of standing heat tends to be short. Insemination the following day is satisfactory for cows seen bulling in the afternoon or late at night. Farmers who practise 'do-it-yourself' AI can, of course, manage the time of insemination with greater precision and may, with good technique, increase conception rate above 60%.

Pregnancy diagnosis

The veterinary student spends many hours with his or her hand in a cow's rectum learning to recognise the structure of the ovaries by palpation and to diagnose the existence and stage of pregnancy. This is an art best learnt by experience and outside the scope of this book. Very briefly, the diagnosis of pregnancy may be based on one or more of the following:

(1) Non-return to oestrus 21 days after insemination.
(2) High progesterone concentration in milk 24 days after insemination.
(3) Palpation of the developing foetus and foetal membranes *per rectum* at least 35 days after insemination.
(4) Ultrasound scanning using a rectal probe at least 42 days after insemination.

Non-return to oestrus is, on its own, a very unreliable indicator of pregnancy for a number of reasons. For instance, the cow may have been incorrectly diagnosed as bulling, inseminated at the wrong time of the cycle or while in anoestrus. In either case she would not be seen on heat 21 days later. The milk progesterone assay at 24 days is an excellent way of identifying the cow that is *not* pregnant but false positives can occur if fertilisation has not taken place but life of the corpus luteum is unusually prolonged, if there is early death of the embryo, perhaps because of a genetic error prior to implantation, or if there is a failure of implantation.

Diagnosis of pregnancy by rectal palpation is not really possible until the 35th day of pregnancy. Thereafter it gets progressively easier as pregnancy progresses. It should not be attempted earlier since it is possible to abort a cow when performing a rectal examination during the critical stage of implantation.

The detection of pregnancy in cattle by the use of an ultrasound rectal probe is accurate but not intrinsically any easier than manual palpation nor can it be carried out any earlier in pregnancy. Pregnancy diagnosis by ultrasound has been an enormous boon to sheep farmers. For cattle, however, it appears to be a rather expensive gadget designed to provide a solution to a non-existent problem.

The importance of pregnancy diagnosis is to ensure that cows which have been inseminated and not subsequently seen bulling are, in fact, pregnant. Table 12.1 showed that the difference between a herd with a good and poor fertility record can exceed £100 per cow per annum. I have already emphasised the fact that the most common reason for poor fertility and an extended calving interval is poor heat detection. The most common reason for the mistaken assumption that a cow is pregnant is that she was served at the wrong time in the first place and not seen when bulling subsequently. Thus the poorer the heat detection, the greater the need for accurate pregnancy diagnosis. It is also possible that the cow has some disorder of the ovaries or uterus that requires veterinary treatment. In either case it is vital that the farmer diagnoses the non-pregnant cow as soon as possible and takes appropriate action to get her back in calf as soon as possible.

Rebreeding

Assuming all goes well, the cow is inseminated, conceives and carries her calf to term, then delivers it without complications or subsequent infection. The next question becomes, 'How soon should she be rebred?'.

A healthy cow will normally recommence follicular development and ovulate for the first time 15 to 25 days after calving. In about 50% of cows this first ovulation is accompanied by a silent heat, i.e. she displays no sexual activity nor sexual attraction to another cow or bull in the days prior to ovulation. The second ovulation, at 36–46 days is normally accompanied by a standing heat. The first heat should be recorded but there is little point in inseminating cows under 50 days post-calving since conception rates are less than 50%, probably because in some cows the uterus has not recovered to a degree when it will permit implantation. Cows should be inseminated at the first oestrus that occurs more than 50 days after calving. If conception rate is 60% then:

	Cows served	Cows conceiving	Mean calving to conception interval
First service	100	60	60
Second service	40	24	81
Third service	16	10	102

and so on, giving an average calving interval of 354 days.

Infertility

The failure of a cow to breed can be due either to mismanagement or to some physiological disorder, possibly attributable to mismanagement or to infectious diesease. This section deals with physiological disorders and disease. It is, however, essential at the outset to restate the normal position illustrated by Table 12.3. Even in the best managed herds heat detection is only 80% accurate and conception rate following a single service by AI only 60%, so that only 48% (60% × 80%) of cows ovulating within a 21-day period are likely to conceive. The majority of herds are not managed this well and an unsatisfactory calving interval, in excess of 390 days, can usually be improved by attention to the relatively simple matter of heat detection and timing of insemination. Nevertheless there is a great deal that can go wrong with the reproductive tract of an individual cow.

Ovarian dysfunction

The three main failures of normal ovarian function are:

(1) *Anoestrus*. The ovary becomes quiescent. There is no follicular development, no ovulation and cycling stops.
(2) *Follicular cysts*. One or more mature follicles fail to rupture, and enlarge further into a cyst that secretes oestrogen.
(3) *Luteal cysts*. A ruptured follicle or (more probably) a mature, unruptured follicular cyst, luteinises and persistently secretes progesterone.

A few cows and rather more heifers may exhibit true anoestrus with no follicular development for more than 60 days after calving. Other cows will commence oestrous cycles within the first 30 days after calving and then stop again. The most usual cause is inadequate nutrition and excessive loss of body condition following calving. I shall discuss nutrition and fertility in more detail later.

About 4% of cows develop follicular cysts. The reasons for this are unclear. The condition appears to be heritable. In the UK it is allegedly a greater problem in Jerseys than in Friesians. In Sweden, the national incidence has been reduced from 10% to 5% by rejecting bulls for AI if from a family of cows with a history of cystic ovaries. It has also been suggested that management stresses may contribute to cystic ovaries but this is unproven.

A cow with follicular cysts displays erratic and usually excessive sexual behaviour. In most cases she comes on heat at irregular but abnormally short intervals, maybe 8–12 days apart. This heat may last 3–4 days. Furthermore, the cow may be almost permanently excited by the presence of other bulling cows in the herd. This condition is accurately described as nymphomania. Affected cows are infertile and rapidly take on a very ragged appearance, not only because of the amount of riding that they give and take but because the excessive secretion of oestrogen relaxes the ligaments of their pelvis and alters the shape of the caudal spine, elevating the tailhead.

Cows with persistent follicular cysts become progressively masculinised due to secretion of androgens, and display bull-like behaviour, rearing, pawing the ground and generally creating a nuisance. A nymphomaniac cow can, in theory, be used as an aid to heat detection but, in practice, she is usually too much bother. A few cows with follicular cysts show no sexual behaviour at all.

About 70% of ovarian cysts are follicular. The other 30% are luteal cysts which secrete progesterone. Most of these probably start life as unruptured follicular cysts. It is not easy to differentiate between follicular and luteal cysts by rectal palpation but they can be clearly distinguished using the milk progesterone assay (Ball, 1983). Treatment of follicular cysts usually involves an injection of luteinising hormone to rupture the follicle followed by a further treatment with progesterone (e.g. using a PRID) in an attempt to re-establish normal oestrous cycles. The prognosis is fair. The rarer luteal cysts can usually be cured by an infection of prostaglandin.

Uterine disorders

In normal calving, the calf is ejected from the sterile environment of the uterus, the epithelium of the placenta separates from the uterine tissue at the cotyledons, and the afterbirth is expelled shortly afterwards. The uterine muscle and endometrium at this time weigh about 8 kg. There follows a period of rapid loss of uterine tissue such that its weight falls to about 2.5 kg by Day 7 and 1 kg by Day 25. Phagocytes invade the tissues in massive numbers to clear up the cellular debris, some of which is absorbed directly into the blood and lymph and some of which passes into the body of the uterus and is expelled through the vagina. Despite the large amount of cell destruction and the involvement of immune cells, this is normally a sterile or aseptic process, i.e. no infectious organisms are involved. However, the natural process of involution can be complicated by infection in the following circumstances:

(1) Abortion caused by organisms which attack the placenta, such as *Brucella abortus* or *Campylobacter*.
(2) Introduction of infection through the vagina when calving without assistance but in an infected environment.
(3) Tissue damage and introduction of infection during an assisted calving.

Any of these can cause complications which impede the normal processes of uterine involution and may lead to retained placenta and/ or endometritis (inflammation of the uterus).

Retained placenta

Most cows expel the afterbirth within 6 hours of calving. A few may wait 24 hours but still expel the whole afterbirth normally, in a single piece. Any cow that has not shed the afterbirth by 24 hours will then probably not shed it normally for several days until its cotelydon attachments have putrefied. Some cows appear to tolerate the internal consequences of this increasingly stinking mess hanging from the vulva without showing any signs of ill-health. Others can become sick quite rapidly.

   The condition is caused by a failure of the placental epithelium to separate from the uterine wall at the cotyledons. This may be due to infection and adhesions caused, for example by *Brucella abortus*, or by inadequate shrinking and separation of the uterine tissues under the influence of oxytocin at the time of calving. In practice, the factors that predispose to retained placenta are:

(1) Infectious abortions.
(2) Premature calvings, including those induced by drugs such as prostaglandin.
(3) Prolonged calvings caused by dystocia or twins.
(4) Infection acquired at calving.
(5) Milk fever.
(6) Nutritional disorders (possibly – see below).

When cases of retained placenta occur sporadically (about 5% of calvings) it is often difficult to establish which, if any, of these factors was responsible. An incidence greater than 10% is clear evidence that some management factor is at fault and needs to be corrected.

There is considerable dispute as to the best form of treatment for retained placenta. Formerly vets would routinely attempt to remove the placenta manually, then insert a pessary or irrigate the uterus with an antiseptic or antibiotic. Nowadays the vet may inject oestrogen or oxytocin in an attempt to induce a natural, aseptic separation of the placenta. Others may use antibiotic treatment but make no attempt to remove the placenta unless the cow is clearly ill. On balance, manual removal appears to do internal harm more often than the 'hands off' approach, so long as the cow is treated with antibiotics (see de Bois, p.479, in Krag and Schallenberger, 1982). Whether the retained placenta is manually removed or not, affected cows are nearly always slow to return to oestrus and even then conception rates are usually only about 40%. In de Bois' study the calving-to-conception interval in cows with retained placenta was on average 108 days for those that did conceive and 15% of cows were culled for infertility.

Endometritis

In practice, endometritis is almost always associated with infection acquired at the time of calving and organisms involved are often the pus-forming inhabitants of dirty cowsheds, *Corynebacterium pyogenes* and *Fusobacterium necrophorum*. The ability of the natural defence mechanisms of the cow to overcome such infection is impaired if the cow is malnourished or suffers a metabolic disorder. In severe cases the cow may be very ill indeed with fever, anorexia, shock and a foul-smelling, brown discharge from the vulva. Such animals require immediate veterinary attention and a few die before any treatment has time to work. More commonly, endometritis proceeds to a chronic stage in which the cow is quite well but continually discharges a mixture of white pus and mucus from the uterus into the vagina. This condition, 'the whites', can, if untreated, persist for many weeks and seriously delay rebreeding. Conventional treatment with antibiotics is now augmented by prostaglandin injection. This directly contracts the uterus, so expelling some of the pus, but also induces follicle development and the secretion of oestrogen which stimulates phagocytosis and other local defence mechanisms.

Once again, the main consequence of endometritis is a prolonged calving-to-conception interval and once again, the main cause of endometritis is poor hygiene at calving, in particular the insertion of an unscrubbed, non-sterile arm into the vagina.

Early foetal death

The probability of a successful fertilisation following insemination at the right time of oestrus is about 90%, yet only about 60% of such cows are found to be pregnant when examined after 40 days. About 8% of fertilised ova will fail to develop because of some genetic defect in (or incompatibility between) sperm and ovum. However, most cases of early foetal death occur during the time of implantation at 25–30 days of pregnancy. In some cases the tiny developing embryo may be aborted, unnoticed. More often, the foetus and its membranes are absorbed into the uterine tissues. This can inhibit the secretion of luteolysin (the natural prostaglandin) and cause the corpus luteum to persist so that the cow may not return to oestrus for many weeks after death of the embryo. This is a common cause of infertility or of cows returning to oestrus perhaps 60–80 days after an apparently successful insemination. The condition can be treated simply by injection of prostaglandin. However, it is sometimes difficult to distinguish between the presence of a live foetus or a dead, resorbing foetus by rectal examination; both conditions will give high milk progesterone levels, and prostaglandin, administered by mistake to a pregnant cow, will induce abortion.

Abortion

Abortion means, in common parlance, the abnormally premature expulsion of a foetus either dead or in no fit state to live. In fact the distinction between abortion and early foetal death is arbitrary. Some foetuses that die at 25 days are aborted, others that die at a much later stage are not aborted (or at least not at once) but are retained in the uterus where they mummify.

In areas of the world where brucellosis is endemic, this is the main cause of infectious abortion. Since eradication of brucellosis from the UK dairy herd, the number of abortions reported to the Ministry of Agriculture has been about 2% of all calvings. Farmers have a legal obligation to report all abortions so the real figure is probably not much greater than this. Despite the fact that the Ministry of Agriculture were unable to establish an infectious cause in three-quarters of the cases submitted to them in 1983, it is probable that most cases of abortion in mid to late pregnancy are due to infection. Physical causes such as acute nutritional stress, transport or manhandling, can induce abortion but are not, I think, responsible for many cases in practice. Brucellosis was discussed in Chapter 10. Sporadic cases of abortion can arise in association with summer mastitis caused by *C. pyogenes* or acute salmonellosis. Abortion can become endemic in herds infected with IBR virus or *Leptospira* (the hardjo variety). This condition can also cause mastitis with a particularly severe loss of milk and can infect people who come into direct contact with cattle, causing a chronic debilitating disease having symptoms rather similar to brucellosis. It is possible to control both IBR and leptospirosis by vaccination. In these, as in all cases of abortion, it is not only a legal necessity to seek veterinary advice, it also makes good sense.

Nutrition and infertility

Cows in poor condition are harder to get back in calf. This is obvious to every dairy farmer with high-yielding cows 'milking off their backs' and even more obvious to beef farmers whose cows are producing less but are grossly short of food. Malnourished cows not only lose fat but also protein in muscle and skin. A standard procedure is adopted for condition scoring cows according to the amount of subcutaneous fat (and muscle) in the loin area around the lumbar vertebrae and in the area between the pelvis and the tailhead. The appearance of the tailhead in cows with condition scores from 0 to 5 is illustrated in Fig. 12.3. The scoring is as follows:

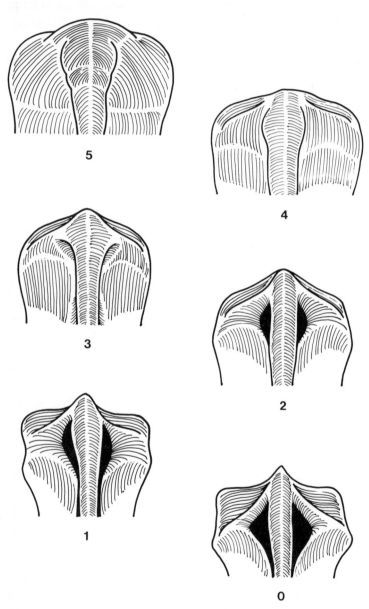

Fig. 12.3 Condition scoring of dairy cows from the appearance of the tailhead. 5 = grossly fat, 4 = fat, 3 = good, 2 = moderate, 1 = poor, 0 = very poor. (For further information, see text.)

5. *Grossly fat:* tailhead buried in fatty tissue, pelvis impalpable even with firm pressure.
4. *Fat:* folds and patches of soft fatty tissue under the skin. Pelvis palpable with firm pressure but transverse processes of lumbar vertebrae impalpable.
3. *Good:* all bones are palpable but well covered with fatty tissue.
2. *Moderate:* all bones easily palpable, muscle concave around tailhead, some fatty tissue.
1. *Poor:* muscles and tailhead and lumbar vertebrae shrunken and concave. No fatty tissue palpable. However, skin is supple and freely moveable.
0. *Very poor:* emaciated. Skin is thin, tight, with no tissue palpable between skin and bone.

It is generally agreed that cows should have an average condition score between 2.5 and 3.5 at the time of calving. Lower scores mean less milk. Cows with condition scores above 3 are prone to the ketosis/fatty liver complex which not only compromises their ability to achieve peak yields but *may* predispose to retained placenta and cystic ovaries (Chapter 7).

It is thus important for cows not to be too fat at calving. However, the high yielding adult cow is expected to lose 30–60 kg body weight during the first 80 days of lactation which corresponds to a drop of 1 to $1\frac{1}{2}$ points in condition score. In heifers a loss of 15 kg is equivalent to a 1-point drop in condition score. When average condition score falls below 2, fewer cows come into oestrus and conception rate in those that do is only about 50%.

The farmer or veterinary surgeon facing a hard problem of poor fertility presenting as an abnormally extended calving-to-conception interval can only work from good records. These are described in detail by Esslemont, Bailie and Cooper (1985). If they can establish that the problem is not primarily one of poor heat detection and then eliminate those individuals known to have had specific problems such as endometritis or cystic ovaries, the most probable cause of delayed conception is improper nutrition. Feeding strategies designed to ensure that cows are neither too fat at calving nor too thin when due for rebreeding were discussed in Chapter 6. I shall not recapitulate them here, but offer a list of questions designed to establish what, if any, aspect of nutrition is responsible for the poor fertility and how it may be cured.

(1) *Condition score at calving?*
Is average condition score satisfactory (2.5–3.5)? If cows are, on average, fatter than this, is there any evidence that this is causing problems, e.g. ketosis, poor appetite, or other possible complications such as retained placenta? If not, then this is probably not the source of the problem. Are cows too thin at calving? This can often happen in autumn-calving UK herds after a dry summer with insufficient grass. If a high-yielding cow is too thin at calving she is likely to be much too thin by breeding time.

(2) *Condition score of cows due for breeding?*
A comparison of condition scores of cows due to calve in early, mid and late lactation will reveal how accurately nutrient intake is meeting nutrient requirement at each stage of lactation. Cows, for example, on a two-step flat rate feeding strategy could well be too thin at 80 days and too fat at 300 days. Cows on a fairly ruthless brinkmanship strategy may be very thin by 80 days.

(3) *Milk yields in early lactation?*
Are milk yields satisfactory or depressed? Low milk yield plus infertility in thin cows is a clear indication that cows are simply not getting enough food. Cows that are milking well but losing weight fast are likely to be receiving a diet that is generously meeting their need for metabolizable protein but which is deficient in metabolizable energy, probably because they cannot eat enough. If possible the cows should be encouraged to eat more out of parlour using, for example, a highly palatable, relatively low protein food like sugar beet pulp. If this is not possible it may be necessary to substitute a concentrate ration with a lower protein concentration (e.g. 16% for 18%) and accept a small drop in milk yields.

(4) *How well does nutrient supply match requirement?*
I repeat that nutritional infertility is usually caused by an absolute shortage of nutrients – the most important being energy or a relative deficiency of energy with respect to protein. This can be checked by comparing energy and protein supply with requirement as described in Chapter 5 and possibly taking a mini metabolic profile (blood concentrations of glucose, ketones and urea). There are in addition literally thousands of references in scientific journals to problems of infertility associated with practically every other major and minor nutrient, the most popular being the minerals phosphorus, copper and selenium. There may be specific effects of some of these nutrients on specific aspects of fertility. For instance, treatment with vitamin E and selenium *may*

be of benefit in a herd with abnormally high incidence of retained placenta. However, most dairy cows are generously provided with minerals and vitavins in their compound rations so have no difficulty meeting requirements within the constraints of dry matter intake. Moreover, if a deficiency of a specific nutrient (e.g. copper or phosphorus) does occur its effect on fertility is likely to be indirect: a slowing down of metabolism leading to a reduction in appetite. It is possible to imagine (although I have never seen) a situation where dairy cows are losing condition, milking poorly, not displaying oestrus and with a poor appetite for readily available energy and protein-rich forage and concentrate. In such circumstances it would be correct to include specific mineral deficiency in the differential diagnosis along with other possibilities such as parasitism or poisoning, but in the realistic expectation that the obvious solution is more often than not the correct one.

Artificial manipulation of fertility

All the procedures described in this chapter so far have been directed towards maintaining normal fertility in the working dairy cow, i.e. one calf per year. In recent years, however, a range of scientific procedures has been developed which makes it possible to manipulate, more or less radically, the processes of reproduction with the aim of improving quality or quantity defined, respectively, by genetic improvement and increased calf numbers (Roche, 1989; Woolliams and Wilmut, 1989). My review of these procedures will, as always, contain within the definition of a quality, quality of life as perceived by the cow.

The ovaries of the new-born heifer calf contain thousands of primordial follicles, each of which is potentially capable of maturing and releasing an egg. In real life, however, the dairy cow will only release 20–40 eggs in her lifetime and give birth to four to seven calves. The animal breeder will argue entirely fairly that conventional breeding (I take artificial insemination to be conventional) severely restricts the potential of the superior cow to transmit that superiority to her offspring via multiple ovulation and embryo transfer (MOET). In the previous chapter I quoted evidence to show that the rate of genetic progress achievable by MOET is little greater than that from a good progeny testing programme which casts serious doubts on MOET as a cost-effective method for sustaining genetic progress in the national dairy herd, even before welfare considerations are taken into account. The most attractive use for MOET is in international trade; shipment of genetically superior embryos is preferable to movement of live cattle on grounds of cost, health and welfare.

Multiple ovulation is achieved by injecting cows with follicle-stimulating hormone. This will cause. on average, 12–18 follicles to develop to the point of ovulation. These can now be recovered by a 'non-surgical' method which involves passing a cannula through the cervix containing a two- or three-way catheter which permits injection of fluid into the uterus and recovery of fertilised ova. These ova are then inserted into the uterus of a recipient cow once again via the cervix. This process of 'inovulation' is carried out six days after oestrus. In many ways it is similar to artificial insemination except that (1) the catheter is passed more deeply into the uterus; (2) penetration of the cervix is more difficult on day six than during oestrus. Skilled operators can usually complete the process within two minutes. There is a dispute as to whether epidural anaesthesia should be used routinely. This involves injecting local anaesthetic into the spinal canal at the tail head. The case for epidural anaesthesia is that it eliminates any pain involved in the procedure and incidentally makes life easier for the operator. A cynic might also add that it ensures the job can only be done under a veterinary supervisor. The case against is that the procedure does not normally appear to distress the cow and anaesthesia stops the cow kicking out when it does hurt and so increases the risk that the operator may do damage. I accept both arguments and believe that the decision should be left to the judgement of the operator (provided welfare is properly taken into account) rather than imposed by legislation.

An alternative approach to MOET is to recover developing oocytes from the ovary, either by laparoscopy (which involves a very small incision in the abdominal wall) or at slaughter. These may be matured and fertilised *in vitro*. One commercial argument for this has been to try to make twinning a routine procedure in (especially) beef cattle. At first sight the strictly commercial advantages of twinning appear obvious – twice as many calves for sale to offset the costs of maintaining their mothers. In practice, however, twinning is a very suspect procedure indeed. Measured strictly in terms of number of twins born the results are disappointing. When other factors are taken into account the alleged 'benefits' of twinning actually become negative. Two of my former students, Carys David and Owen Davies, in association with Roger Eddy, one of our top cow vets, carried out an economic analysis of the consequences of (spontaneous) twinning in dairy cattle. Their results, based on 503 twin calvings and matched comparisons from an

Table 12.4 Financial consequences of spontaneous twinning in dairy cows

	Single calf	Twin calves
Income		
milk sales, gross profit margin	612.12	640.32
sale of calves	137.17	184.86
sale of cull cows	89.25	148.75
	838.54	973.82
Expenditure		
calving – conception above 85d	34.20	105.45
serves/conception	13.60	17.60
replacement cows	138.60	231.00
ratio heifers:cows	153.03	160.08
less calves from replacement heifers	50.84	50.75
Vet costs	1.84	8.92
	371.72	573.80
Income – expenditure	466.82	400.02

overall population of approximately 20,000 cows, are summarised in Table 12.4.

Cows bearing twins obviously earned more money from the sale of calves. In this example twin calves were valued at £109 each and calf mortality was 15.2%, single calves at £145 with mortality 5.4%. Mothers of twins also gave more milk than their matched pairs for reasons I cannot explain. There was no difference in the incidence of problems at calving. However, twinning seriously increased the problem of infertility, increased both the time taken to get cows back in calf and the number of cows culled. When all major factors, including the consequences of increased ratio of heifers to cows in the herd are taken into account, every case of twins lost the farmer £66.8 relative to the normal delivery of a single calf.

The welfare problems of embryo transfer do not end at ovulation. Twinning poses problems after, rather than at calving. Transfer of embryos of large cattle (e.g. Charolais, Belgian Blue) into small recipient cows potentially increases the risk of obstetric problems due to an overlarge calf. This fear should not, however, be exaggerated since the size of the calf at birth and its ease of passage through the pelvic canal are determined less by the genotype of the calf and more by the size, shape and condition of the mother.

A new problem has arisen very recently following the process of cloning calves by transfer of nuclear material into enucleated oocytes ('nuclear transfer'). These calves are nearly always larger than normal calves at birth, the average being about 25% larger but some calves are born as 'giants'. The incidence of Caesarian sections for calves cloned by nuclear transfer has been 20%. The duration of pregnancy is not prolonged and these calves grow normally after birth. I don't know what the cloners have done in their attempt to produce increased numbers of identical superior animals and neither, as yet, do they.

Part V

Coda

13   A Cow For All Seasons

During a period of approximately fifty years from the 1930s to the 1980s, agricultural productivity in the developed world expanded faster than at any other time in history. This expansion owed as much, if not more, to political will as it did to new science. Decisions to give the farmer guaranteed prices for food and free access to government-sponsored research and development looked a good idea in the depression of the 'dirty thirties', an even better idea during the Second World War and still a reasonable idea during the post-war campaign to produce 'food from our own resources'. Whenever the incentive to produce more was strong, farmers (including dairy farmers) responded magnificently with the eminently foreseeable consequence that supply came to exceed demand from those in the developed world who could afford to buy. At the same time guaranteed sales and guaranteed high prices within Europe and North America effectively squeezed out free and fair trading with the Third World, both imports and exports.

Farmers within Europe and North America cannot be blamed for gearing their production to suit the rules of the game as laid down by politicians but few could now deny that support systems for agriculture have been getting seriously out of hand. The absurdities of overproduction are well enough known: mountains of cereal, beef and skimmed milk guarded at great expense in a fortress Europe while the majority of the world goes hungry. Some of the excesses of agricultural support are equally bizarre. In 1975 there was one government adviser to UK agriculture for every 700 hectares of farm land. In the USA, apparently every dairy cow currently costs the taxpayer $700 in government grants which implies that it would have been cheaper for the US government to offer a free cow to anyone who wanted one.

The dairy industry has benefited from support for both milk and beef production but has, within Europe, paid an artificially high price for cereals. The imposition of quotas on milk output per farm has been reasonably successful at balancing overall supply to demand. It also calls for some rather subtle manipulation of nutrient supply on the individual farm to maximise grow profit margins within the constraints of the quota in any one year. However, one early, rather romantic view that quotas would encourage extensification with cows eating much less concentrate and working less hard to produce less milk per head has not come to pass. As dairy farmers become more exposed to the forces of the free market they are forced to compete ever more fiercely with one another to stay in business. The weak, or weary, sell up and the successful buy over their quota.

A free market in dairy production (with or without quotas) is all very well up to a point since the consumer benefits and so, too, does the more successful farmer. There are, however, aspects of dairy farming which cannot simply be assessed in terms of the short-term rules of the free market but are absolutely vital if we are to preserve quality of life for both man and those animals that serve man. The three key issues are, of course, sustainability, pollution and animal welfare.

We live in interesting times. It is clear that the conspicuous consumption of the developed world cannot be sustained. It is just to assume that the poor in the developing world have the same rights to creature comforts as ourselves yet the very unsustainability of our lifestyle makes it unexportable. Meantime, throughout much of the world, especially sub-Saharan Africa, things are getting worse. Furthermore, our understanding of those animals that serve us has (or should have) evolved to the point where we can no longer think of them as mere machines for our own use but as sentient creatures who can perceive quality in life within a spectrum ranging from suffering to pleasure. As I wrote at the outset, the greater the extent of our dominion over the life of any sentient animal the greater becomes our responsibility to provide it with a reasonable quality of life and a gentle death.

Where then does this leave the farmer and his dairy cows?

Farming, ethics and public opinion

The image of the farmer is not what it was. Once upon a time (to use that most beguiling of invitations to fantasy) the farmer was the jolly red-faced provider, or the choleric old gentleman who chased you with a stick for stealing apples but, in either case, was a worthy and visible member of the community. Those who bought the food could look over the fence and see the cows and chickens. It was not perfect but it was life, proceeding in full public view. One of the great problems of intensive agriculture is that it has divorced the consumer from the farm. The great majority of the public have withdrawn from the realities of working a living from the land and retreated deep into their caves, forming their impressions from the flickering images of reality on the cave walls (*The Republic, Book 6:* Plato really understood the power of television). Images are very powerful and invariably incomplete. Reality is usually untidy, complicated and less eye-catching. As the consumer becomes more urbanised his or her choice of food becomes ever more susceptible to image manipulation. The producer, on the other hand, has no choice but to live with reality.

The extreme expression of consumer resistance to animal produce is not just the personal decision to avoid them (the vegan position) but the demand that man should cease to exploit farm animals altogether and 'set them free'. One flaw in this argument is that having set them free into this crowded world, there is almost nowhere for them to go. In Chapter 1 I explained in some detail how ruminants occupy a different but complementary ecological niche to man, relying largely on grasses, which we cannot digest, and by-products, i.e. the parts of plants which we choose not to eat. The good farmer works the land using the plants and animals most suited to that land, expends time, money and effort to maintain and increase the production and health of his crops and herds, and earns his living by harvesting food and other products from them *at the right time.*

Most farm animals are killed while in their prime – dairy cows are an exception. If we accept that farm animals neither know in advance the number of their days nor have long-term ambitions (except the protection of their offspring for the time they are together), it becomes impossible to argue that humane, instantaneous slaughter causes more distress than a slow decline, through increasing decrepitude, to a natural death in the wild. If we can accept this then it is in no way brutal to expect farm animals to earn their keep by contributing milk, eggs or meat in return for their board and lodging.

If we do not milk, kill or otherwise exploit farm animals we can no longer afford to keep them, except for a few survivors in zoos and farm museums. The vast majority of cattle in the UK set free with no supplementary food would starve to death during the winter. Cattle can overwinter without supplementary food in UK conditions but only in small numbers, in certain environments, and they get very hungry indeed. Having substituted the sin of neglect for the sin of exploitation and seen the cattle and sheep disappear from our hills, meadows and parklands, we would then be faced with the interesting ecological problem of 'who cuts the grass?'.

The good husbandman accepts the necessity to kill his animals as a fact of life but also recognises their capacity for suffering and therefore their right to be treated in life and at the point of death with humanity and compassion.

The other common criticism of modern intensive livestock farming is that animals are considered only as machines or units of resource within an enterprise whose sole objective is to make a profit. This is less easy to rebut although easier for dairymen than for most livestock farmers since it is difficult to imagine a group of people who can tie up so much capital and then work so hard for so long to guarantee such a small percentage return. Nevertheless it is right that dairy farmers and those, like the veterinary profession, who serve dairy farmers, should give serious attention to this criticism, partly to acknowledge public concern and so maintain the sympathy of the customer, but mainly to ensure that the animals in our charge are also in our care.

We must ask ourselves three questions:

(1) To what extent has the intensification of dairy farming in the interests of increased productivity had harmful or beneficial effects on the welfare of the dairy cow?

(2) How might the welfare of the dairy cow be improved without impairing her economic efficiency?

(3) Are there any unacceptable aspects of modern dairy husbandry that cannot be eliminated without economic cost so that if change is to be made it must be enforced by law?

Whether we consider animal welfare in absolute terms or as an exercise in public relations, it cannot be divorced from economics. By this I do not simply mean the costs of production but also the elasticity of demand from a heterogenous population of consumers whose concern for animal welfare has to compete with more powerful forces such as the concept of healthy eating and the necessity to live within a budget.

In a free market the dairy and livestock industries have no option but to provide the type of food that the consumer wishes to buy. Keeping it cheap undoubtedly helps but not at the expense of all else. If the use of additives such as hormones to increase the efficiency of milk and meat production actually reduces sales because a significant proportion of the public likes to think of their meat and milk as additive-free then hormones are not a good idea. There is no point building a better mousetrap in a world that worships mice.

The public image of animal welfare is too big an issue to review here in detail. Most criticism relates to intensification – battery hens, sows in stalls, veal calves in crates – and is based on the argument that the animals are denied the freedom to express their natural behaviour. Within this limited definition of welfare, dairy farming has largely escaped criticism, but a more comprehensive analysis of the welfare of the dairy cow according to all five freedoms (Chapter 1 and below) does generate some serious worries.

In tackling the question 'Has intensification caused welfare to become better or worse?' it is necessary to distinguish between and balance up two opposing trends. Increasing the number of livestock managed by any one man tends to be detrimental to welfare by reducing the economic importance of each individual animal. On the other hand, intensification has been made possible only through improved scientific understanding of the requirements of each species for health and efficient production and this, properly applied, must be good for the animals. The peasant farmer took care of his only cow because she was his most prized possession. He cared for her to the best of his ability and if they starved, through poverty, ignorance or oppression, they starved together. The chicken farmer with a million laying birds in cages has probably achieved a very low incidence of infectious disease and death in his birds because it is economic so to do. However, one dead bird would not cause him to lose any sleep and one bird feeling 'off colour' and depressed would go unnoticed.

This is perhaps a fanciful example but the conflict between economics and welfare can be very real. Suppose the poultry farmer discovered that by feeding Diet B instead of Diet A he could achieve an improved gross profit margin, notwithstanding a 2% increase in mortality due to bone fractures, he would almost certainly opt for Diet B. He would equally certainly explore ways of reducing mortality on Diet B so as to increase profit margin further. Here the conflict between welfare and economics exists in the short term but the producer works to resolve the conflict.

The problem of extreme confinement in intensive units, however, cannot be resolved in this way. If one farmer puts four and then five birds in a cage originally designed for three, and the birds do not die nor stop laying eggs, then his competitors have no option but to do the same if they want to stay in business. This is an example of an unacceptable aspect of husbandry that cannot be eliminated without economic cost and so calls for new legislation.

I have, up till now, deliberately avoided using examples from the dairy industry to illustrate the problems that we face in deciding how far we should go in an attempt to meet animal 'rights'. We cannot be sure of the ways in which intensification of the dairy industry may have done good or harm to the welfare of the cows because we do not have adequate records of malnutrition, discomfort, injury and disease in the days before intensification. It is perhaps more useful to summarise and recapitulate some of the ways in which current dairy husbandry falls short of the absolute standards set by the 'five freedoms' and consider how they may be improved.

The five freedoms

(1) Freedom from hunger and thirst

Improvements in our understanding of digestive physiology and nutrient requirements, coupled with improved methods for conserving winter food, mean that cows are less likely to suffer from seasonal malnutrition than formerly in well-managed, productive herds which generate enough income to pay for winter food. When poverty strikes anywhere in the world, man and beast suffer together. While dairy farming remains profitable, nutritional problems are likely to be those of excess, e.g. acidosis and laminitis associated with large amounts of starchy concentrates. If economics force dairy farmers towards more extensive, low-input, low-output systems, the incidence of some production diseases may decrease, to be replaced by other specific forms of malnutrition, e.g. mineral deficiencies or, more likely, energy deficiency, i.e. chronic hunger.

(2) Freedom from prolonged discomfort

Selection of dairy cows for increased yield has produced big, angular animals with large udders and crooked hind legs and so exacerbated the problems of standing, walking and especially lying on hard surfaces. The MAFF Welfare Code for Cattle (1983) recommends that 'dairy cows should not be kept on a totally slatted area. A solid floored area incorporating straw or a suitable bed should be provided to ensure comfort and reduce the risk of injury.' The most important criterion of a suitable bed is a hygienic material with the 'give' of a mattress

(Chapter 8). Since this cannot be justified on economic grounds alone it becomes a 'right' that can only be guaranteed by legislation.

(3) Freedom from pain, injury and disease

Dairy farmers should really require no incentive to minimise injury and disease because sickness and death cost money. Although mastitis and, to a lesser extent, endometritis, remain as perennial problems, there can be little doubt that the incidence of most infectious diseases in dairy cattle has declined over the last 40 years. Cattle are, on the whole, healthier, so by this criterion their welfare has improved. The most glaring exception to this rule is foot lameness, especially due to white line disease, laminitis and solar ulcer, which has, almost certainly, got worse. The reasons for this are complex (Chapter 8) but can be attributed to changes in conformation, nutrition and housing, all engineered deliberately in the interest of improved productivity and profit. The dairy farmer is in exactly the same position as my imaginary poultry farmer who consciously elects to feed Diet B in the knowledge that he will have more lame birds. The solution to those aspects of lameness that can be attributed to nutrition and housing on a particular farm is a matter of professional advice and good stockmanship and not a matter for legislation. The problem of conformation is more stubborn. The modern dairy cow does not have a satisfactory shape. Selection towards a more dual purpose animal (dairy and beef) would almost certainly reduce the incidence of hind foot lameness but I think it most unlikely that this will happen either in response to market forces or the force of law.

An alternative solution to the problem of the over-distended udder that distorts the hind legs and throws abnormal stresses on the outer claws of the hind feet is to milk the cows more often. I shall return to this thought in the section entitled *Manipulation of the environment*.

(4) Behavioural freedom

This was discussed in Chapter 8. It is achieved largely by a combination of good building design and good stockmanship and is not a matter for legislation.

(5) Freedom from fear and distress

The greatest cause for concern is the treatment of cows at the end of their life, especially those animals that are injured, sick or paralysed and so cannot walk onto a lorry and be taken away in reasonable comfort. The law recognises that, except in an emergency, no such animal should be transported if this causes unnecessary suffering. Emergencies are, in fact very rare and the description is designed to cover eventualities such as the movement of a downer cow from a field to a box where she may be better treated. Any farmer in possession of an animal which is suffering pain or distress that is likely to endure has an obligation to relieve its condition by obtaining the services of a veterinary surgeon or by slaughtering the animal humanely using the services of a slaughterman or knacker. *De jure*, all such animals should be killed on the farm. *De facto*, there is no way of ensuring that this happens. A cow that is injured or sick can only be accepted at an abattoir if it is accompanied by a casualty certificate stating that in his or her opinion the carcass will be fit for human consumption, wholly or in part. The certificate is concerned only with meat hygiene, not animal welfare. In 1986 the British Veterinary Association produced a new standard form of casualty certificate which includes a section (which has no power in law) to indicate whether the animal should be killed on the farm or, failing that, how far it should be permitted to travel. It would be a great boon to cattle welfare if this section were made mandatory so that no farmer could send, nor abattoir accept, a casualty animal without a veterinary certificate to declare that it was fit to be moved without causing unecessary suffering.

Economics of production

The welfare of the dairy cow and the dairy farmer can be ensured only so long as the farm remains economic. As a general rule (there are exceptions) the farmer who makes the most money from dairying has the healthiest cows. What is more, he can afford to improve overall standards of comfort and hygiene and care for the sick. It is easier to be good when one has money.

Dairy farmers are constantly advised to improve profit margins by improving physical performance in terms of such things as margin over concentrates, calving interval, season of calving, age of heifers at first calving, etc. I have discussed the physical principles that underlie these at some length already and do not intend to repeat them here. It is, however, salutary to consider the extent to which profit margins for a dairy farmer are *not* related to the physical performance of his cows. An analysis of profit margins in the top and bottom 25% of herds recorded by Genus Management in 1991 is summarised very briefly in Table 13.1.

This table is inevitably unrepresentative because it excludes the poorest farmers scratching along without benefit of records. It also does not examine variation, and therefore alternative strategies within the top 25%. The most interesting observation to my mind is the relatively small variation between top and bottom herds in milk yield and concentrate use. The most impressive difference (27%) in terms of physical performance is in calculated milk yield from forage. The

Table 13.1   Some factors affecting profitability in dairy herds in 1991 (Genus Management)

	Top 25%	Bottom 25%
Physical performance per annum		
milk yield (l/cow)	5790	5317
concentrate use (kg/cow)	1490	1472
milk from forage (l/cow)	2339	1841
Economic performance		
milk and calf sales	1178	1082
less herd replacements	52	54
gross output	1126	1028
variable costs, food	322	314
sundries	95	114
Gross margin (£/cow)	709	600
(£/ha)	1538	1164
Fixed costs (£/ha)		
wages	166	207
power and machinery	178	264
miscellaneous	324	355
interest	94	249
total fixed costs (£/ha)	762	1075
Net profit (£/ha)	+776	+89

greatest differences between top and bottom herds are seen within fixed costs: paid labour (i.e. that not done by the farmer and his family), power and machinery and especially interest due on borrowed capital. The message of this table is not quite 'Don't borrow money' but any dairy farmer wishing to invest money to improve some aspect of his dairy unit should first carry out or commission a detailed business plan for his existing and planned enterprise to establish whether the projected returns will justify the initial and continuing cost of his new investment.

Table 13.1 tells us little about the effect of different feeding and breeding strategies on gross profit margins. Table 13.2, also from Genus (1992) examines in a little more detail factors associated with variation between farms in milk yield, concentrate use and stocking rate. Comparing low (<5400 l/cow/year), medium (5400–6200 l/cow/year) and high (>6200 l/cow/year) yielding herds reveals, as expected, a substantial increase in concentrate consumption, gross margins over purchased foods (MOPF, £/cow) and margins over purchased food and fertiliser (MOPFF, £/ha). However, ranking farms according to concentrate use (kg/cow) reveals a much smaller increase in MOPF (£/cow) although the expected increase in MOPFF (£/ha) as the ratio of purchased to home-grown food increases

Table 13.2 Analysis of factors affecting profit margins and UME per hectare in costed herds (Genus, 1992)

Independent variable	Milk yield (l/cow)	Concentrates (kg/cow)	UME (GJ/ha)	MOPF (£/cow)	MOPFF (£/ha)
Milk yield (l/cow)					
<5400	(4924)	1182	83	767	1934
5400–6200	(5781)	1463	88	886	1934
>6200	(6639)	1848	91	986	2268
Concentrate use (kg/cow)					
<1350	5304	(1086)	90	842	1786
1350–1850	5840	(1581)	87	878	1945
>1850	6364	(2152)	80	906	2127
Stocking rate (LSU/ha)					
<2.25	5599	1387	(75)	858	1606
2.26–2.75	5782	1502	(94)	880	2076
>2.75	5747	1571	(111)	859	2535

MOPF = gross margin over purchased feeds
MOPFF = gross margin over purchased feeds and fertilisers
UME   = utilised metabolizable energy

Table 13.3 Performance and profits in six top dairy herds

	A	B	C	D	E	F
Milk yield (l/cow)	8784	7430	6264	6222	7739	8096
Concentrate use (kg/cow)	1737	2530	959	824	3247	3247
Stocking rate (cow/ha)	2.93	2.35	1.83	2.51	1.98	2.72
MOPF £/cow	1062	1020	1014	1008	991	1017
MOPFF £/ha	3017	2279	1081	2451	1868	2723

accompanied by a decrease in utilised ME (UME) from grass. Increasing stocking rate was largely unrelated to milk yield, concentrate use and MOPF (£/cow) but was the most effective contributor to profit when expressed as MOPFF (£/ha). The implication of this is that maximum gross profit (in UK conditions) is achieved by maximising UME from grass or other home-grown food, so long as milk output is not allowed to fall below quota.

It is probably better not to over-generalise about the factors that contribute to gross profit margin in dairy herds. I have always been fascinated to observe the extreme variation that exists between individual top farms. Table 13.3 selects six of the top twelve farms recorded by the Milk Marketing Board in 1989.

All these herds are very impressive. Herd A is exceptional by any account, milk yield, stocking rate, MOPF and MOPFF. Herds B–F are all rather similar in terms of MOPF (£/cow) despite a range of concentrate use from 824–3247 kg/cow. Herd D is, to my mind, particularly impressive, achieving a MOPFF (£/ha) of £2451 or less than 1 tonne concentrates per cow and at a stocking rate of 2.5l/ha.

I cite these examples not as an essay in economics but as an illustration of the flexibility of the dairy cow and her ability to generate similar profits (per head) across a wide range of systems of feeding and management. However hard times may be, this should give cause for optimism. The dairy cow is not yet so specialised that she cannot adapt to change.

Economics of scale

Figure 13.1 illustrates the economics of scale for UK dairy units. The numbers are based on 1985 prices but the shape of the response curves for profit margins are essentially the same today. Fixed costs are high, over £400 per cow for herds of under 40 cows, but decline to a minimum value of £280 per cow at 100 cows. Variable costs increase slightly with increasing herd size (partly because cows in bigger herds tend to be given more purchased food) but approach a maximum of £480 at a herd size of 150.

When herd size is between 100 and 150 cows, total costs are minimal and gross returns per cow reach their maximum; in short, there is maximum economy of scale. A herd size of 100–150 is within the capacity of one milkman milking twice daily and is, moreover, the size of enterprise that can be accommodated within the larger family farm with a relatively high input of unpaid labour and, probably, a relatively small debt to the bank. The economics of dairying may not be rosy but it does seem, in the UK at least, that the better family farm is likely to be as efficient as the mega unit.

The dairy cow of tomorrow

Throughout this book – until now – I have tried to convey my understanding of the dairy cow only in the light of existing knowledge.

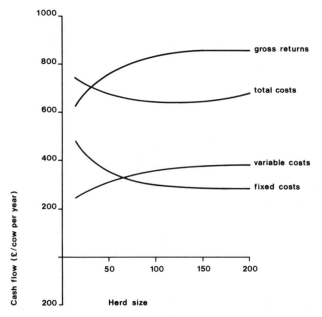

Fig. 13.1 Effect of herd size on output, costs and gross returns (£/cow per year, from Seabrook [1985]).

Where I have used phrases like 'I think' it has been to express personal doubt rather than dogma. You will, I trust, permit me at last a little free speculation as to what we might expect of the dairy cow in the future. I shall assume, as always, that the only sure thing one can say about economic forces is that they will continue both to fluctuate and to dominate the pattern of progress, almost irrespective of what may be shown to be biologically possible or deemed to be ethically acceptable.

In a strictly economic sense the objectives are to increase yield, to improve (or otherwise manipulate) milk composition and to improve the efficiency of food conversion to milk. Within the constraints imposed by economics we must also aim to improve the health and welfare of the cow and work towards a more sustainable form of food production. To achieve these objectives we can manipulate the cow, her feeding or her environment. I shall consider each of these three options in turn.

Manipulating the cow

The conventional way to manipulate the appearance or performance of an animal is by controlled breeding, i.e. choice of the donors of both sperm and eggs to favour the development of certain desired characteristics. Modern high-tech aids to breeding such as multiple ovulation and embryo transfer (MOET, Chapter 11) and *in vitro* fertilisation do not alter the basic principles of fusion of sperm and egg and differ only in degree from the well established practice of artificial insemination. MOET is merely a process for acceleration of the rate of genetic change.

Performance may also be manipulated by regular injections of substances such as bovine growth hormone or somatotropin (BST) which alter the genetically determined metabolic processes of the cow. Finally, the cow may be manipulated by genetic engineering. If genetic material is injected into the early embryo it will, with luck, become part of the cow's overall genetic identity and be transmitted to subsequent generations. Alternatively a new gene may be introduced to an animal after birth, perhaps on a carrier virus and modify some feature of the animal, say the immune system, for its life only. It can be argued that the first deliberate exercise in genetic engineering was performed by Dr Jenner when he infected his son with cowpox (*Vaccinia*) thereby protecting him from the greater scourge of smallpox.

When attempting to evaluate methods of manipulating cows for profit it is, I think, helpful to concentrate on how the procedure will be perceived by the cow and what good it may do for society (or for the cow herself), rather than become embroiled in a discussion as to whether or not any specific procedure is 'unnatural'.

The Farm Animal Welfare Council (1988) has defined the circumstances in which manipulation of an animal may lead to a significant deterioration in welfare. These include 'the manipulation of body size, shape or reproductive capacity by breeding, nutrition, hormone therapy or gene insertion in such a way as to reduce mobility, increase the risk of injury, metabolic disease, psychological distress, skeletal or obstetric problems and perinatal mortality'. Clearly the double-muscled Belgian Blue, achieved by conventional breeding, offends by this definition as did the giant Beltsville pig produced by genetically engineering it to produce excessive quantities of growth hormone (see Wheale and McNally, 1990).

The arguments for and against bovine somatotropin (BST) are rather less clear cut. This compound, almost but not quite identical to natural bovine growth hormone has been genetically engineered into bacteria so that it can be produced relatively cheaply and in large quantities. It has to be administered to cattle by injections, at regular intervals since its active life is short and it is degraded within the digestive tract. It augments the action of endogenous growth hormone and can increase milk yield by anything from 5 to 30% depending on the nutritional state of the animal, at any stage of lactation. Increasing milk yield obviously imposes increasing nutrient demands on the cow who responds (when she can) by increasing food intake so as to establish a new equilibrium between input and output about three weeks after the onset of 'treatment' with BST.

There is no doubt that BST works in the sense that it can increase milk yield. Moreover, although it does not increase the net efficiency of milk synthesis (k_l) it can increase overall efficiency of food utilisation by increasing output relative to maintenance cost. There is a huge amount of published material on BST and, I am bound to say, most of it reveals no significant difference between treated and untreated animals in production disorders.

There is a tendency for fertility to fall when BST administration commences before cows are back in calf and there are some reports of slight increases in mastitis and lameness. It is, however, most unscientific to give undue emphasis to selected reports that reinforce one's own prejudices. Nevertheless I shall continue to oppose BST mainly because the most important welfare problems for the dairy cow are the stresses of overwork (Fig. 7.2) and BST makes cows work harder. Breeding a cow to produce more milk involves more or less subtle changes in the whole animal – one of the less subtle ones being the production of a larger cow. Increasing yield through improved nutrition, provided it creates no digestive disorders, involves moving the cow towards her genetic potential. The regular use of exogenous hormones inevitably involves driving an animal faster than it is designed to run. I wrote earlier that all forms of manipulation to an animal should be subjected to cost-benefit analysis. With BST the costs are those of regular injections and the possibility (not the certainty) of an increased risk in production disease. The 'benefits' amount to an enforced increase in milk yield which the consumer in developed society does not want, the poor in the Third World cannot use, which no dairy farmer actually needs and which will drive some out of business. I rest my case.

The most promising scope for genetic engineering lies in manipulation of the composition of milk. Considering, for the moment, milk only as a source of food, there are large differences both within and between species in the concentration of the main constituents in milk, these differences being determined by genes. For example,

| | Concentrations, g/kg | | | 'Price' |
	Protein	Fat	Lactose	Pence/litre
Friesian cow	33	39	45	17.1
Jersey cow	37	49	46	20.6
Sheep	48	62	65	26.5
Horse	20	11	48	8.6
Bitch	81	98	35	41.0

As I have written earlier, the most simple way to manipulate the composition of milk products is after the milk leaves the cow, to produce butter, cheese, low-fat yoghurt, ghee, etc. Nevertheless lactose is usually produced surplus to requirement for protein and fats and a low-lactose milk would have added value. Jersey milk has more protein and fat relative to lactose than Friesian milk. Sheep milk is more concentrated than cow's milk but the proportions of the three major constituents are similar which suggests that engineering genes for milk synthesis in sheep into cattle would not carry much commercial advantage. The most interesting example cited above is bitch's milk which is not only rich in protein and fat but has a protein lactose ratio of 2.3:1 compared to 0.73:1 for Friesian cattle. At current prices for cow's milk, bitch's milk would fetch 41p/litre! This example shows that the ratio protein:lactose is subject to genetic manipulation. If the genes for milk protein synthesis in the dog (or some similar, less emotionally sensitive species) were sufficiently few in number to be located and inserted into cattle it might be possible to engineer a cow that produced milk with a composition more suited to consumer demand. Of course, it might just taste dreadful.

A much more promising role for genetic engineering is the insertion of genes which will code for the production in milk of specific proteins which can be used for the treatment of disease in man. Many of the inherited disorders of mankind such as haemophilia or cystic fibrosis are caused by simple genetic defects that prevent affected individuals from synthesising one or more specific proteins essential for life. A far greater proportion of the population may have a genetic predisposition to certain infectious diseases or cancers. There is undoubtedly an exciting future for prevention and treatment of disease based upon genetically engineered versions of proteins essential for health which we lack wholly or in part.

The most likely source of these proteins will be cultures of genetically engineered microorganisms. However, the AFRC Roslin Institute at Edinburgh has succeeded in engineering essential proteins for the treatment of haemophilia and cystic fibrosis into sheep in such a way that it can be recovered from their milk (Woolliams and Wilmut, 1989). These transgenic animals can, of course, transmit this ability to their offspring.

This advance in genetic engineering was revealed to the public at about the same time as BST. The response, in general, has, I believe, been extremely level-headed. Engineering a human gene into sheep could have aroused paranoia because (to those who think that way) it involves the mixing of human life with that of the lower animals. However, it has not, presumably because the use of animals to produce a pharmaceutical essential for a proper life in sick children is such a thoroughly worthwhile idea. On the other hand, BST has aroused massive consumer resistance, not because the science is particularly bizarre but because it is so obviously unnecessary.

A final welfare note: like it or not, the care given to an individual animal is governed very much by the value we put on it, be it the cash value of a thoroughbred stallion or the emotional value of a pet dog.

Such animals may have care lavished upon them but, at the other end of the spectrum, the spent battery hen at the end of lay is worth almost nothing and is treated accordingly. Cattle or sheep that have been genetically engineered to produce a vital and expensive pharmaceutical will acquire the cash value of racehorses and be treated accordingly. I am quite sure that any cow, if she could be made to understand the argument, would give serious attention to the prospect of serving mankind in other ways than as a source of food.

Manipulation of food supply

There have been great advances in the feeding of dairy cows in the last twenty years not because any new food materials have suddenly been discovered but because we have achieved a far better understanding of how to supply an optimal balance of nutrients first to the micro-organisms in the rumen and then to the cow. Where the main food source has been fresh or conserved grass, the concentrate supplement has changed radically from being seen largely as a source of extra energy, usually based on starch, to become increasingly a source of extra protein. Where grass continues to be the cheapest source of nutrients, I predict that the energy source in concentrate feeds will continue to be based on digestible fibre, not only on grounds of cost but also because it will ensure efficient fermentation. It will be important to improve our ability to characterise foodstuffs in terms of their ability to yield quickly and slowly fermentable energy, quickly and slowly degradable crude protein, and undegraded, metabolizable energy and protein. It would in theory be desirable to define nutrient supply even more precisely than this, e.g. in terms of specific VFAs and amino acids. This degree of refinement already exists in pig and poultry nutrition. It is, however, a long way from being sufficiently precise to be of practical use in ruminant feeding. This is because (1) we do not know with sufficient precision the contribution of VFAs and microbial protein to ME and MP; (2) when the contribution of undegraded amino acids is small their amino acid composition has to depart substantially from the optimum if it is to have a perceptible effect on the overall efficiency of protein use.

I believe therefore that the greatest advances in nutrition will come from a better understanding of the food materials that already exist. There are two other approaches: (1) to improve techniques for conserving crops, especially grass and whole-crop cereals; (2) to manipulate the rumen population so as to improve the rate and extent of (especially) fibre digestion. I stated in Chapter 6 that I believe that our approach to ensiling grass is not quite right and whole-crop cereals (other than maize) a long way from right. There is considerable interest at present in fundamental research designed to increase the cellulolytic activity of microorganisms in the rumen and even to produce fermentors of lignin. Such organisms have been created already and shown to work under laboratory conditions. Whether or not they could survive the fierce competition in the rumen remains to be seen. It would perhaps be more logical to use (say) a microorganism that degraded lignin outside the rumen to improve the digestibility of hay or straw in forage.

This book has inevitably been biased towards cattle production in UK conditions where grass is the main source of food and cellulose is the main substrate for rumen fermentation. In many warmer and drier areas cereals (especially corn or maize) form the main food source and therefore starch becomes the major substrate. Reduced cereal prices in the UK and the possibility of global warming could make grass a less attractive crop for us too. High starch diets need to be handled with care to avoid problems of ruminal indigestion and acidosis, but if they are they can promote very high yields. In North America, high corn starch diets for cows are regularly supplemented with buffers (e.g. sodium bicarbonate), yeast-based compounds and other sources of B vitamins (e.g. niacin or nicotinamide). Such food additives have obvious attractions to those who wish to sell them and, under conditions of high-starch feeding, some of them sometimes improve performance. The antibiotic food additive, monensin, has had a major effect on the efficiency of growth in beef cattle by manipulating rumen fermentation in such a way as to reduce both energy losses as methane and the risk of acidosis. In theory, an antibiotic that did not leave the digestive tract could be used to manipulate fermentation in a way conducive to improved efficiency of milk synthesis.

The promoters of food additives, like all salesmen, need to be viewed with a cool eye. The fact that a buffer, or a yeast supplement has been shown to work in certain circumstances does not mean that it will work equally well (if at all) where circumstances are different. The basic principles I adopt when investigating all such claims are as follows:

(1) The capacity of the mammary gland to secrete milk usually exceeds the capacity of the digestive tract to supply nutrients so

that anything which increases nutrient supply has the potential to increase milk yield. This applies particularly when the energy source is mostly in slowly fermentable form and intake is constrained by gut fill.

(2) When the diet consists mostly of slowly fermentable energy, rumen fermentation is slow but stable and is unlikely to be enhanced by the addition of buffers, B vitamins, etc.

(3) When the diet is rich in quickly fermentable energy (starch) the potential to supply nutrients is high but this may be constrained by disorders of rumen fermentation (acidosis, etc.). In these circumstances food additives which restore normal rumen function may well enhance performance.

To summarise this argument in a single sentence, it is easier to improve performance with a food additive if you have impaired it first with improper feeding.

Manipulating the environment

Perhaps the most unnatural thing about the dairy cow is that her environment and her working day have been so distorted to accommodate man's conception of what constitutes a working day and a working environment. The most unnatural of all unnatural acts is to breed a cow capable of producing over 40 litres/day then restrict her to two milkings instead of the 5–7 feeds she would normally allow her calf. This must be a major contributor to udder distension and thereby to mastitis and lameness.

Other aspects of her life dictated more by our whims than hers include two rushed meals of highly-fermentable feed during the few minutes she is in the parlour, interpretation of her sexual behaviour by a species which does not really understand the message which she is trying to convey, treatment of disease with antibiotics and other drugs to counter organisms which are identified by microbiologists, but seldom (unlike us) medicines and nursing to counter the symptoms of sickness. I could go on.

The point I wish to make is that whereas most aspects of dairy husbandry are satisfactory they have largely involved adapting the cow to systems of our own devising. If one starts from the other end, namely a fundamental understanding of the physiology and behaviour of the cow, it is possible to envisage husbandry systems which bear little relation to those which we accept as normal. It is also possible to challenge that most sacred of beliefs, namely 'stockmanship'. I do not deny for a second the importance of stockmanship or the differences between a good and bad stockman but may I suggest that our conventional view of stockmanship is entirely paternalistic: the better the stockman the more he or she does for the cows. It is just possible that the cow (or, to give a more obvious example, the sow in a stall) might prefer to make a greater and more constructive contribution to the quality of her own life. This immediately raises the prospect of the cow-operated service station, incorporating a computer and robotics that would:

(1) recognise the cow on entry;
(2) dispense food appropriate to her needs;
(3) wash the udder and attach the milking machine;
(4) diagnose early signs of mastitis through the presence of clots, changes in osmotic pressure or increased temperature of the udder. If mastitis is suspected the milk would be directed towards a dump line and the cow directed into an enclosure where she can be examined and treated if necessary by the stockman and/or veterinary surgeon.
(5) analyse milk for progesterone concentration, store the analysis from successive days and so predict with precision the time of oestrus. Once again, cows in oestrus would be directed out of the parlour into an area where they can be retained for insemination.
(6) record milk yield, remove the clusters automatically and disinfect the teats;
(7) record body temperature, body weight and food intake, identify any cow showing abnormalities of any of these three and warn the stockman that the cow may be in the early stages of infection or metabolic disease;
(8) identify the presence of ketone bodies in milk or saliva.

If such a system could be made to work reliably it would not only free the stockman from the more repetitive chores of the working day, thereby giving him more time to observe his cows; it would be better suited to the physiological needs (and passing whims) of the dairy cow. It would also never lose its temper.

Such systems are already being developed. Each component of the system is already feasible and not outrageously expensive. The most pressing need now is to undertake some market research with the cows themselves to discover just what it is they want, i.e. what will motivate them to enter the service station regularly to ensure that they are milked (say) five times a day in early lactation and at least twice daily from 200 days onwards. It may be that the autumn-born cow at grass in early summer will dry herself off because she is insufficiently motivated to enter the stall if the only incentives are milking and a little concentrate food. These questions are currently uppermost in my mind and I hope to have some answers by the third edition. Meantime, I repeat our approach to this problem, like our approach to most problems of animal management and animal welfare, is to ask the animal what it thinks. Devising useful questions in such a way that animals can understand them is sometimes tricky but always rewarding.

Cows for ever

Ultimately the future of the dairy cow will depend upon her ability to contribute to sustainable agriculture. There is no doubt that cows in the developed world have consumed a lot of food that could, in theory, have been fed directly to man but, as I pointed out in Table 1.4, the milch cow has the potential to produce 60–70% more protein and energy for man than she consumes in the form of food that we can use. She is therefore truly able to share the resources of the earth with us. For most of the people in the world, the solution is not and is not likely to become the high-tech, robotics-based option being developed in Europe and North America but a highly sustainable, low cost approach where the greatest advances are likely to come from investment in education rather than in capital goods. Crotty (1980) has conducted a detailed economic analysis of cattle production in the developing world, revealing where, on occasions, things have gone spectacularly wrong but also where cattle are and will remain vital to sustained development. I shall not attempt to summarise his complex arguments at this late stage but instead quote Michael Halse (1975) of the National Dairy Development Board of India:

'It is meaningless to say that Indian farmers should not divert land to grass fodder for milk production, because that is not what happens. India's draught and milch animals are sustained mainly on herbage which would otherwise be wasted. In a successful dairy project a farmer gets one or two better animals. He increases milk production per animal by 50–80 litres and *increases his income by as much as 30–50%*. He does not 'divert land' from food production.

He simply makes his farm more productive. Modern dairy technology can bring rurally produced milk economically to the cities and urban consumer – rupees can be directed back to the rural milk sheds through the dairy co-operatives for investment in improved production.'

The gap between this sort of farming and the high-yielding Holstein herds of the western world (with or without robots) is immense. However, both approaches have shown progress, albeit sometimes at the expense of animal welfare and the environment. Moreover, both can continue to progress in a way that is friendly to both earth and cow.

If we examine the reasons for this progress with some humility we discover that relatively little can be attributed to things we have changed.

We can grow more grass and cereals (if we invest more capital energy in the form of fossil fuels) but the quality of grass, corn and barley is much the same as it was. We can increase milk yield (slowly) by controlled breeding but the average modern Holstein is still no better than the fabulous Magdalena in 1810. We can produce an effective vaccine against a serious but straightforward disease like foot-and-mouth, but still must control mastitis through attention to good hygiene. We can introduce genes for the production of novel proteins by microbes and by the mammary gland but we have failed to devise a bio-industrial process for converting fibre into food that is as adaptable, mobile, self-replicating, low cost and utterly endearing as the cow.

The most impressive evidence for real progress has come from the unspectacular, but scientifically sound, pursuit of the perennial complex problems of nutrient supply, hygiene, control of lameness and fertility. Such patient work seldom produces the 'Gee Whiz' effect induced by topics such as *in vitro* fertilisation or genetic engineering. It does, however, tend to provoke more satisfactory answers to the necessary follow-up question, 'So what?'.

I conclude therefore as I began. Our objective is to improve production and health in dairy cattle in a way that is consistent with sustainability and quality of life. If this is the end, what are the means? Understanding the dairy cow.

Further Reading

The publications listed below include references cited in the text and other books and articles not previously quoted in any specific context but which have been of value to me and will be of equal value to readers seeking further information.

Part I   How the cow works

Allen, D. (1990) *Planned Beef Production and Marketing.* 3rd ed. Blackwell Scientific Publications, Oxford.

Blaxter, K.L. (1967) *The Energy Metabolism of Ruminants.* 2nd ed. Hutchinson, London.

Blaxter, K.L. and Webster, A.J.F. (1991) Animal production and food: real problems and paranoia. *Animal Production* **53**, 261–70.

Blowey, R.W. (1985) *A Veterinary Book for Dairy Farmers.* Farming Press, Ipswich.

Burnet, M. and White, D.O. (1972) *Natural History of Infectious Disease.* Cambridge University Press.

Castle, M.E. and Watkins, P. (1984) *Modern Milk Production; its Principles and Applications for Students and Farmers.* Faber, London.

Church, D.C. (1979) *Digestive Physiology and Nutrition of Ruminants. Volume 1, Physiology; Volume 2, Nutrition.* O & B Books Ltd, Corvallis, Oregon, USA.

Committee on Medical Aspects of Food Policy (1984) *Diet and Cardiovascular Disease.* HMSO, London.

Curtis, S.E. (1983) *Environmental Management in Animal Agriculture.* Iowa State University Press, Iowa, USA.

Esslemont, R.J., Bailie, J.H. and Cooper, M.J. (1985) *Fertility Management in Dairy Cattle.* Collins, London.

Farm Animal Welfare Council (1983) *Code of Recommendations for the Welfare of Livestock: Cattle.* Ministry of Agriculture, Fisheries and Food. Leaflet 701.

Fraser, A.F. and Broom, D.M. (1990) *Farm Animal Behaviour and Welfare*, 3rd ed. Baillière Tindall, London.
Girardier, L. and Stock, M.J. (1983) *Mammalian Thermogenesis*. Chapman and Hall, London.
Greenhalgh, P.R., McCallum, F.J. and Weaver, A.D. (1981) *Lameness in Cattle*. 2nd ed. Wright Scientechnica, Bristol.
Hafez, E.S.E. (1968) *Adaptation of Domestic Animals*. Lea & Febiger, Philadelphia.
Hafez, E.S.E. (1980) *Reproduction in Farm Animals*. 4th ed. Lea & Febiger, Philadelphia.
Hill, W.G., Thompson, R. and Woolliams, J.A. (1990) *Proceedings 4th World Congress on Genetics as applied to Livestock Production*. Vol. XIV. Edinburgh University Press.
Kilkenny, J.B. and Herbert, W.A. (1976) *Rearing Replacements for Beef and Dairy Herds*. Milk Marketing Board, Thames Ditton, Surrey.
Kleiber, M. (1961) *The Fire of Life*. Wiley, New York.
Kon, S.K. and Cowie, A.T. (1961) *Milk, the Mammary Gland and its Secretion*. Vols. I, II. Academic Press, New York.
Leach, G. (1976) *Energy and Food Production*. APC Science and Technology Press, Guildford.
Leaver, J.D. (1983) *Milk Production, Science and Practice*. Longman, London.
Lister, D. (1984) *In Vivo Measurement of Body Composition in Meat Animals*. Elsevier, London.
McDonald, L.E. (1980) *Veterinary Endocrinology and Reproduction*. Lea & Febiger, Philadelphia.
McDonald, P., Edwards, R.A. and Greenhalgh, J.F.D. (1988) *Animal Nutrition*. 4th ed. Oliver & Boyd, Edinburgh.
McFarlane, M.V. (1976) Water and electrolytes in domestic animals. In: *Veterinary Physiology* (ed. J.W. Phillips), pp. 461–539. Wright Scientechnica, Bristol.
Milk Marketing Board (1986) *UK Dairy Facts and Figures*. Thames Ditton.
Monteith, J. and Mount, L.E. (1974) *Heat Loss from Animals and Man*. Butterworths, London.
Ørskov, E.R. (1981) *Protein Nutrition in Ruminants*. Academic Press, London.
Pimentel, D. and Pimentel, M. (1979) *Food, Energy and Society*. Arnold, London.

Rook, J.A.F. and Thomas, P.C. (1983) *Nutritional Physiology of Farm Animals*. Longman, London.

Tizard, I. (1982) *An Introduction to Veterinary Immunology*. Saunders, London.

Van Soest, P.J. (1982) *Nutritional Ecology of the Ruminant*. O & B Books Ltd, Corvallis, Oregon, USA.

Webster, A.J.F. (1981) Weather and infectious disease in cattle. *Veterinary Record* **108**, 183–187.

Webster, A.J.F. (1985) *Calf Husbandry, Health and Welfare*. Collins, London.

Wood, P.D.P. (1976) Algebraic models of the lactation curves for milk, fat and protein production with estimates of seasonal variations. *Animal Production* **22**, 35–40.

Part II   Feeding the dairy cow

Agricultural Research Council (1980) *The Nutrient Requirements of Ruminant Livestock*. 2nd ed. Commonwealth Agriculture Bureau, Slough.

Blood, D.C. and Radostits, O.M. (1989) *Veterinary Medicine*. 6th ed. Baillière Tindall, London.

Blowey, R.W. (1985) *A Veterinary Book for Dairy Farmers*. Farming Press, Ipswich.

Broster, W.H., Johnson, C.L. and Tayler, J.C. (1980). *Feeding Strategies for Dairy Cows*. Agricultural Research Council, London.

Broster, W.H., Phipps, R.H. and Johnson, C.L. (1985) Principles and practice of feeding dairy cows. *Technical Bulletin 8*. National Institute for Research in Dairying, Reading.

Department of Agriculture and Fisheries for Scotland. Reports I–IV of the Feedingstuffs Evaluation Unit, Rowett Research Institute, Aberdeen.

Ekesbo, I. (1968) *Disease Incidence in Tied and Loose-housed Dairy Cattle*. Acta Agriculturae Scandinavia Supplt. 15.

Holmes, W. (1989) *Grass; its Production and Utilisation*. 2nd ed. Blackwell Scientific Publications, Oxford.

Interdepartmental Working Party (1991) *Report on the Protein Requirements of Ruminants*. Agricultural Development and Advisory Services. HMSO, London.

Leaver, J.D. (1983) *Milk Production, Science and Practice*. Longman, Harlow.

McDonald, P., Edwards, R.A. and Greenhalgh, J.F. (1988) *Animal Nutrition*. 4th ed. Oliver & Boyd, Edinburgh.

McDonald, P., Henderson, A.R. and Heron, S.J.E. (1991) *The Biochemistry of Silage*. Chalcombe Publications, Marlow.

Ministry of Agriculture, Fisheries and Food (1978) *Complete Diet Feeding of Dairy Cows, Second Year Report*. MAFF Publs., Alnwick, Northumberland.

Ministry of Agriculture, Fisheries and Food (1980) *Nutrient Allowances and Composition of Feedingstuffs for Ruminants*. Booklet 2087. MAFF Publs. Alnwick, Northumberland.

Ministry of Agriculture, Fisheries and Food (1984) *Silage Additives Used in England and Wales*. MAFF Publs. Alnwick, Northumberland.

National Research Council, USA (1985) *Nutrient Requirements of Domestic Animals, No. 3, Dairy Cattle*. National Academy of Sciences, Washington, USA.

Owen, J. (1979) *Complete Diets for Cattle and Sheep*. Farming Press, Ipswich.

Reeve, A., Baker, R.D. and Hodson, R.G. (1986) 'The response of January/February calving British Friesian cows to level of protein supplementation'. *Animal Production* **42**, 435.

Underwood, E.J. (1980) *Mineral Nutrition of Livestock*. 2nd ed. Commonwealth Agriculture Bureau, Slough.

Van Adrichen, P.W.M. (1977) Editor, *Third International Conference on Production Disease in Farm Animals*. Centre for Agriculture Publishing, Wageningen, Netherlands.

Webster, A.J.F. (1985) *Calf Husbandry, Health and Welfare*. Collins, London.

Webster, A.J.F. (1992) The Metabolizable Protein System for Ruminants. In: *Recent Advances in Animal Nutrition* (1992) (eds. J. Wiseman and D.J.A. Cole), pp. 93–112. Butterworths, London.

Wilesmith, J.W., Ryan, J.B.M. and Atkinson, M.J. (1991) Bovine spongiform encephalopathy: epidemiological studies on the origin. *Veterinary Record* **128**, 199–203.

Part III   Housing, health and management

Baggott, D. (1982) 'Hoof lameness in dairy cattle.' *In Practice* (supplt. to *Veterinary Record*) **4**, 133.

Blood, D.C. and Radostits, O.M. (1989) *Veterinary Medicine*. 6th ed. Baillière Tindall, London.
Blowey, R.W. (1985) *A Veterinary Book for Dairy Farmers*. Farming Press, Ipswich.
Blowey, R.W. (1986) An assessment of the economic benefits of a mastitis control scheme. *Veterinary Record* **119**, 551.
British Standards Institution (1981) Design of buildings and structures for agriculture *2.2 Livestock Buildings*. BS 5502(2.2) London.
Clarke, P.O. (1980) *Buildings for Milk Production*. Cement and Concrete Association, Slough.
Ekesbo, I. (1968) *Disease Incidence in Tied and Loose-housed Dairy Cattle*. Acta Agriculturae Scandinavia. Supplt. 15.
Grandin, Temple (1980) Observations on cattle behaviour applied to the design of cattle handling facilities. *Applied Animal Ethology* **6**, 19.
Greenhalgh, P.R., MacCallum, F.J. and Weaver, A.D. (1981) *Lameness in Cattle*. 2nd ed. Wright Scientechnica, Bristol.
Ministry of Agriculture, Fisheries and Food, Alnwick, Northumberland. Booklet 2424. *Housing and Management of Young Dairy Cattle*. Booklet 2431. *Design and Management of Cubicles for Dairy Cows*. Booklet 2495. *Cattle Handling*.
Noton, N.H. (1982) *Farm Buildings*. College of Estate Management, Reading.
Scottish Farm Buildings Investigation Unit (1985) *Annual Farm Building Cost Guide*. SFBIU, Aberdeen.
Theil, C.C. and Dodd, F.H. (1977) *Machine Milking*. National Institute for Research in Dairying, Reading.
Webster, A.J.F. (1985) *Calf Husbandry, Health and Welfare*. Collins, London.
Wilesmith, J.W., Hoinville, L.T., Ryan, J.B.M. and Sayers, A.R. (1992) Bovine spongiform encephalopathy: aspects of the clinical picture and analyses of possible changes 1986–1990. *Veterinary Record* **130**, 197–201.

Part IV   Breeding and fertility

Allen, D. (1990) *Planned Beef Production and Marketing*. 3rd ed. Blackwell Scientific Publications, Oxford.
Ball, P. (1983) Fertility problems in dairy herds. *In Practice* (Supplt. to *Veterinary Record*) **5**, 189.

Blowey, R.W. (1985) *A Veterinary Book for Dairy Farmers.* Farming Press, Ipswich.

Esslemont, R.J., Bailie, J.H. and Cooper, M.J. (1985) *Fertility Management in Dairy Cattle.* Collins, London.

Falconer, D.S. (1960) *Introduction to Quantitative Genetics.* Oliver & Boyd, London.

Hill, W.G., Thompson, R. and Woolliams, J.A. (1990) *Proceedings 4th World Congress on Genetics as applied to Livestock Production.* Vol. XIV. Edinburgh University Press.

Karg, H. and Schallenberger, E. (1982) 'Factors affecting fertility in the post-partum cow'. *Current Topics in Animal Science and Veterinary Medicine* **20**. Martinus Nijhoff, Amsterdam.

Milk Marketing Board (1986) Thames Ditton, Surrey.
1. *Report of the Breeding and Production Organisation No. 36.*
2. *Bullpower; Premium Friesians-Holsteins 1986–87.*

Ministry of Agriculture, Fisheries and Food (1982) Alnwick, Northumberland.
(a) *An Introduction to Cattle Breeding,*
1. Some theoretical aspects
2. Sire and dam evaluation and selection – dairy cattle.
(b) Leaflet 612. *Condition Scoring of Dairy Cows.*

Nicholas, F. and Smith, C. (1983) Increased rates of genetic change in dairy cattle by embryo transfer and splitting. *Animal Production* **36**, 341–53.

Oldham, J.D., Simms, G. and Marsden, S. (1992) Nutrition-genotype interactions in dairy cattle. In *Recent Advances in Animal Nutrition* (eds. J. Wiseman and D.J.A. Cole). Butterworths, London.

Roche, J.F. (1989) New techniques in hormonal manipulation of cattle production. In *New Techniques in Cattle Production* (ed. C.J.C. Phillips). Butterworths, London.

Webster, A.J.F. (1989) Bioenergetics, bioengineering and growth. *Animal Production* **48**, 249–69.

Willis, M.B. (1991) *Dalton's Introduction to Practical Animal Breeding.* 3rd ed. Blackwell Scientific Publications, Oxford.

Woolliams, J.A. and Wilmut, I. (1989) Embryo manipulation in cattle breeding and production. *Animal Production* **48**, 3–30.

Wray, Naomi R. and Hill, W.G. (1989) Asymptotic rates of response from index selection. *Animal Production* **49**, 217–28.

Coda

Crotty, R. (1980) *Cattle, Economics and Development.* Commonwealth Agricultural Bureau, Slough.

Farm Animal Welfare Council (1983) *Code of Recommendations for the Welfare of Livestock: Cattle.* Ministry of Agriculture, Fisheries and Food, Leaflet 701.

Farm Animal Welfare Council (1988) *Report on Priorities in Animal Welfare, Research and Development.* Farm Animal Welfare Council, Tolworth.

Genus Management (1992) *Genus Farm Business Accounts: Lessons to be learnt from 1991.* Genus, Wrexham.

Halse, M. (1975) Food production and food supply programmes in India. *Proceedings of Nutrition Society* **34**, 173.

Wheale, P. and McNally, Ruth (1990) *The Biorevolution: Cornucopia or Pandora's Box?* Pluto Press, London.

Woolliams, J.A. and Wilmut, I. (1989) Embryo manipulation in cattle breeding and production. *Animal Production* **48**, 3–30.

Index

Figures are shown by page numbers in **bold**